Plants
OF THE
British Isles

First published in Great Britain in 1982 by
William Collins & Co Limited in association with
the British Museum (Natural History)

This edition published in 1986 by
Peerage Books
59 Grosvenor Street
London W1

ISBN 1 85052 051 8

Printed in Portugal by
Oficinas Gráficas ASA

PLANTS
OF THE
BRITISH ISLES

Illustrated by

BARBARA NICHOLSON

Introduction and text by

Frank Brightman

PEERAGE BOOKS

Contents

Barbara Nicholson
A Personal Memoir

By Frank Brightman

Barbara Nicholson, who made all the plant portraits and created the unique paintings that are the subject of this book, was a highly skilled artist and a professional illustrator.

Her early training was at the Southend-on-Sea Municipal School of Art in Essex. Even in those days, in the early 1930s, Southend had become part of a long, continuous strip of seaside town, but the surrounding countryside had not yet succumbed to the great spread of bricks and mortar as it has done so calamitously since. It is a flat country but it appealed to Barbara's sense of landscape. To the north and west was mainly pasture, and to the north-east the marshes provided wide horizons and vistas of sea and sky. Already she had become interested in the wild flowers of the saltings and the farm land.

Although I was not to meet her until many years later, I knew that part of south-east Essex then as she did. I realize now that the masses of buttercups in every field were a consequence of neglectful farming, and an indication that the land was going to pass to other uses. In fact it soon disappeared beneath a vast, unplanned urban sprawl. In later years these unfortunate changes helped to quicken Barbara's interest in nature conservation, even though she was then living in the distant and relatively unspoilt countryside of Dorset.

After beginning her artistic training in Southend, Barbara moved to Oxford. Here she was able to paint a less bleak and more varied countryside. Tentatively at first, she began to develop her interest in the decorative qualities of flowers by sketching them in the Oxford Botanic Garden. After a period of study at the Ashmolean School of Art she went to London and completed her training at the Royal College of Art in South Kensington. Some of the exercises she undertook during this period were anatomical studies executed in the British Museum (Natural History) across the road from the College. Although she was now beginning to turn her attention towards taking up illustration

Barbara Nicholson at her cottage, Carpenters, in the Dorset countryside.

as a profession, her love of the English countryside continued to seek expression in her art; her first paintings to be exhibited at the Royal Academy after she was awarded the ARAC degree were of landscapes in oils.

As a professional illustrator, however, she took up work in the medical field, producing meticulously accurate and at the same time artistically pleasing line and colour drawings of microscope preparations and specimens; this led to her becoming assistant to W. Thorton-Shills, illustrating medical books in collaboration with him. This work was interrupted by the war, when she joined the Women's Auxiliary Air Force, and in 1942 she was commissioned as officer in charge of illustrations for Royal Air Force medical publications in the Air Ministry. After the war she took up her work as a medical illustrator again, although she devoted much of her spare time to making studies of flowers in watercolour. She found herself becoming more and more interested in the structure of plants instead of just their decorative potentialities.

A turning point came in 1957, when Christina Foyle of Foyles Bookshop in London invited Barbara Nicholson to exhibit some of her watercolours at her gallery in the Charing Cross Road. This exhibition was a commercial as well as an artistic success, for sixty-nine of the seventy-two pictures on show were purchased by interested visitors. From this point her botanical work increased and her medical work diminished. She was commissioned by the Oxford University Press to produce all the illustrations for a book that was to become the first of a new series and was published in 1960 as *The Oxford Book of Wild Flowers*. Following the success of this publication, she worked on *The Oxford Book of Garden Flowers*, which appeared in 1964 and was also entirely illustrated with her paintings.

It was at this stage in her career that I first met Barbara. The Oxford University Press, at the suggestion of Professor David Smith, had approached me to write *The Oxford Book of Flowerless Plants*. It was a new departure on their part. The text of *Wild Flowers* and *Garden Flowers* had been written by two or more persons, but a single author was required for the new book, to collaborate closely with the artist, Barbara Nicholson. At our first meeting we were not sure we would get on well together; we both had strong opinions as to the form the book ought to take, and maybe we both put them rather forcefully. Then Barbara invited me to visit her at her home in Dorset. It was an attractive cottage called Carpenters in a quiet village not far from Shaftesbury. The name was abbreviated from the title 'Carpenters' Arms' that it had borne when it had been a small village pub. I found later that in hot weather when the sun beat down on the polished wood floor of the entrance hall a faint beery smell mingled with the odour of beeswax and turpentine. Originally the building had been two cottages, but now it had been converted into a pleasant home and work place. The well-lit studio on the ground floor overlooked a wild garden. Beyond were fields, and over to the east, woodland. We spent an hour or so in the garden. There were unusual and interesting garden plants. There were wild flowers of all kinds, ferns and lichens on the stones in the rock garden and mosses in the crevices between them. We found that though we each looked at plants from a different point of view, our approaches were complementary and so our collaboration began to become a reality.

The first problem concerned the arrangement of our material. *Wild Flowers* had been arranged by flower colour. In *Garden Flowers* the arrangement was even more arbitrary. In *Flowerless Plants* we were not aiming to produce a handbook for we would have room for only a selection of British species. A systematic order therefore seemed inappropriate. So we had little difficulty in agreeing on an ecological approach, arranging the plants by their habitat. We would begin on the seashore and proceed by way of salt marsh and sand dune to grasslands, followed by heaths, mountains, bogs and other wet places and, finally, woodlands. We welcomed the fact that to some extent this would enable us to portray plant communities.

The plants in a community are not only adapted to similar conditions of the environment, but they also have effects on one another. Thus in woodlands the trees shade the undergrowth and the undergrowth shades the plants growing on the ground. Not only is the amount of sunlight reduced, but the moisture of the atmosphere is increased by the presence of the plants. The dead leaves from the trees provide the nutriment for fungi which decompose them and ultimately return minerals to the soil which the trees can use again.

Our arrangement put the sea cliff communities first, in which interaction between the various species is not very great. The most complex of all communities, woodlands, came last. Later, Professor A.R. Clapham in his collaboration with Barbara in connection with *The Oxford Book of Trees* (published in 1975) went very much further than we were able to do in discussing this, and the later book goes into considerable detail about the ecology of woodlands.

The second major problem was the question of names. All plants have scientific names which are useful to know if one is likely to want to find further information about them. Each name consists of two Latin names, the first one acting as a kind of surname. The system was invented by the eighteenth-century Swedish naturalist Carl Linnaeus, and is universally accepted.

In addition to their scientific names, some plants have an English common name going back to very early times. It is interesting that the more conspicuous ones often have more than one common name, each current in a different part of the country. Other plants have an English name invented in the past hundred years or so by writers of flower books. These English names have a different purpose from the scientific names. For instance, as can be seen in the moor, heath and bog paintings in this book, in many of the wilder parts of Britain a conspicuous part is played in the scenery by heather. In many contexts we are not concerned whether this is the true heather (*Calluna vulgaris*) or the so-called bell heather (*Erica cinerea*). In fact heather can mean any of the other species of *Erica* as well; four of them are illustrated in the heath painting. It is possible to coin English names for them all, and in fact it has been done, but there is little point in it. When we wish to be precise we should use the scientific names, and the genuine English common names can be used in other contexts with the meanings they have always held. Anyone who lives in Britain has a clear idea of what is meant by oak; but if we wish to distinguish between the two British species we should say *Quercus robur* and *Quercus petraea*. People living in other parts of the English-speaking world will not be so clear as to the meaning of oak, for in North America, for instance, there are more than a score of species instead of only two. In *Flowerless Plants* we used the scientific names and gave English names as well when these existed. As can be seen in this book, the same practice has been followed for naming flowerless plants in the plant ecology paintings. For flowering plants, an English name is given in every case as well as the scientific name, an invented one being used when no true common one exists.

When we had finished *Flowerless Plants*, we pondered the question of how a pictorial approach to plant ecology could be developed. I conducted some field courses at Kindrogan Field Centre near Pitlochry in Perthshire on the ecology of flowerless plants, and there was no doubt that my students found the illustrations for *Flowerless Plants* very useful and could have made use of more of them. Barbara began to be interested in the seasonal changes that could be seen in the hedgerow in the lane that ran past her cottage and she made a number of preliminary sketches. This was in her spare time only, however, because she was commissioned to work on *The Oxford Book of Food Plants*, published in 1969. All the illustrations, from apples to yams, are her own work, but the text was compiled by three people.

In 1970 I had become Educational Officer at the British Museum (Natural History) and by this time we both felt that Barbara's artistic skills could usefully be employed in explaining ecology in a simple way. It could be done very elegantly, we thought, entirely by reference to flowerless plants.

It soon became clear, though, that the popular audience we wished to speak to would regard this approach as unduly restricted. Another way of looking at ecology might be by considering hedgerows in a fair amount of detail in their various forms and throughout the seasons of the year. This project was very dear to Barbara's heart, but her concept of it was forming and developing only gradually in her mind, and clearly it would be a long-term affair. So why not take hedgerows as just one example amongst a representative selection of natural habitats? There was a need, I was convinced, particularly in secondary schools, for a series of illustrations, poster size, dealing with plant ecology. Some botanical wall charts were currently available but they dealt with the identification of such groups of plants as trees or common weeds and had nothing directly to say about ecology. As far as visual aids were concerned the subject had scarcely been touched upon.

What did we mean by ecology? The word is often used rather vaguely nowadays. As Paul Colinvaux says in his intriguingly entitled book *Why Big Fierce Animals are Rare*,[1] 'Ecology is not the science of pollution, nor is it environmental science. Still less is it a science of doom.' Put very simply, we may say that ecology looks objectively at the ways of life of all living things. In this book we take a narrower view and are concerned with plant ecology. We deal in a very simple way with the relationships between plant communities and their environments. For any individual plant the environment is primarily the place where it lives, which influences the amount of light, water and nutrients from the soil that it receives. Another aspect of the environment is the other plants and also the animals that live in the community.

I put the case for publishing a new kind of wall chart to the Director of the British Museum (Natural History), Dr. (later Sir Frank) Claringbull. He agreed that what I had in mind would make worthwhile Museum publications. We began in a modest way with a series of five which was eventually published in 1972. They were sold as cheaply as possible, and the market aimed at was high schools, colleges and amateur naturalists. From the beginning, however, their appeal proved to be much wider. The general public purchased large numbers of them, first in Britain and then in Europe and in North America. Barbara had succeeded in creating something unique: paintings of great aesthetic charm, combining a set of accurate plant portraits into a whole which at the same time illustrated the plant community as a harmonious unity. The habitat is suggested by skilful use of background features, and each painting has a distinct individuality suiting the mood of the countryside it depicts. The seasonal changes that take place as the year progresses are indicated by placing the spring-flowering species on the left and those that bloom in the summer in the centre. The autumn aspect of the community is shown on the right; sometimes the same plant is shown both in flower and in fruit in the appropriate season. By 1977 fourteen paintings of plant communities, including all the main types to be found in Britain, had been completed. All are reproduced here. In addition, Barbara made a painting of seaweeds and another of culinary herbs.

Although the execution of each painting took about six weeks, much preparatory work was necessary which took considerably longer. Once the list of species to be included had been agreed, Barbara set about getting her preliminary sketches together. Even if she had illustrated a particular plant many times before, she made a point of painting it again from fresh material. And, of course, she had to see all the plants in their habitat in the various seasons of the year. The earlier paintings were of plant communities in southern England, examples of which were to be seen near her home. Chalk grassland was nearby, heathland a little farther away, and the sea coast within easy reach. A little way along the lane from her cottage was a small piece of woodland that she owned, particularly attractive for its wild flowers in the spring. Along the same lane was the hedgerow which she was coming to know very well indeed. Later, in order to study the communities illustrated in the subsequent paintings, she travelled

farther afield. She studied sphagnum bogs in Wales, salt marshes in Essex and fen and fen carr at Wicken Fen in Cambridgeshire. When she came to work on mountain plants and Scottish pine woods she made a special visit to the field centre at Kindrogan in Perthshire. During her stay there the warden, Brian Brookes, showed her the various habitats of the locality and the sites of many interesting plants.

When each painting was very near completion it was closely scrutinized by the staff of the Botany Department of the Museum. Amendments on technical matters were suggested, and it was at this stage that Barbara's professionalism came to the fore. Her whole effort was directed to achieving correctness, and her collaboration with the scientists was happy and successful. However, though a trained botanist can make an accurate description of a plant, this does not necessarily lead to a convincing picture. What was the secret of Barbara's success in conveying the appearance and beauty of wild flowers to ordinary people? I think it was that she had a flair for seeing each plant as a whole, and the skill to capture her vision on paper. She saw the entire plant in the same way that the untrained eye sees it, yet somehow more clearly, so that there is an excitement of recognition in her paintings which convinces us that what she portrays really does look as she shows it.

At the same time as she was working on the wall charts, Barbara was engaged on many other projects. She designed dress materials for Liberty's and wallpapers for Coles', which were more lucrative activities than working for the Museum. She illustrated *The Oxford Book of Trees* previously mentioned, and undertook miscellaneous assignments to produce decorative artwork for other publishers. She was glad to assist the cause of nature conservation by painting plant portraits for the Society for the Promotion of Nature Conservation. And in what time remained she carried on with her especial pleasure of painting the plants in her hedgerow.

The original paintings for the wall charts, done in gouache on card, are at the British Museum

14

(Natural History). The originals of the illustrations for the Oxford books are in various hands, but some of them are at the Hunt Botanical Library, Pittsburgh, U.S.A. In 1977, Barbara was invited to visit the Hunt Library, and the British Museum (Natural History) considered the possibility of arranging with her to combine the trip with preliminary field work leading to a projected series of plant ecology wall charts of American plant communities. However, her doctor strongly advised her not to undertake the stress of transatlantic travel, so she declined the offer of undertaking further wall charts. Barbara Nicholson died in 1978.

An earlier flower painter, one who travelled the whole world to fill her canvases, Marianne North, entitled her autobiography *Recollections of a Happy Life*.[2] Had Barbara Nicholson had the time and the inclination to write her memoirs she might well have chosen a similar title, at any rate as far as her work as a plant illustrator in a much more restricted field was concerned. She knew plants and enjoyed them. She liked to see them in their natural surroundings, and was zealous that plant communities should not be destroyed by thoughtless human activities. I am sad that her somewhat shadowy plan to make a great illustrated book on hedgerows in all their diversity from one part of the country to another and from season to season did not come to fruition, but I suspect that in this project she was content to travel hopefully without really expecting to arrive. Her work, in particular the paintings that follow, has brought pleasure to many thousands of people. Insofar as she succeeded in drawing their attention to the beauty and interest of the wild flowers and other plants of the countryside and the urgent need for us to look after them, she was content.

Frank Brightman

[1] Paul Colinvaux: *Why Big Fierce Animals are Rare: An Ecologist's Perspective*, 1978. Paperback edition, 1981.
[2] Marianne North: *Recollections of a Happy Life*, 1893. Abridged edition, entitled *A Vision of Eden*, 1980.

Author's Note:

The fourteen original paintings by Barbara Nicholson showing the plant communities of different habitats that were published as ecological wall charts by the British Museum (Natural History) are reproduced here, on a reduced scale, with a commentary about the habitats and some information about the plants.

They are arranged with the most important—and complex—climax community, woodland, first. (Woodland is the community that would cover most of Britain were it not for human activities, though comparatively little survives now and what remains is much altered from its natural state.) Next comes pine forest, a Scottish climax community. The mountains of Scotland, Wales and Ireland are too exposed to support the growth of trees, and their vegetation is shown next. Moorland is characteristic of upland areas on flat ground and the gentler slopes. Heath is a somewhat similar community that is found in the lowlands. Moor and heath grade into bog on poor soils where rainfall is high and the ground is not well-drained. In similar poorly drained places where there are more plant nutrients in the soil, fen develops. Open stretches of fresh water have a special community of plants of their own. Wet places by the sea, of course, contain sea salt, and support a characteristic salt marsh community. On sand above the high water mark it is much drier, and quite different plants grow on dunes. The soils of the chalk hills of southern England are well-drained and rather dry; meadows consist of damper grasslands—but both plant communities depend on clearances by farmers and grazing animals. Hedgerows, of course, are entirely man-made, and they have a very interesting community of plants. Human activities also create waste places, and fortunately many plants can survive in some of these too.

More details about these habitats and their communities are given in the commentaries that follow the fourteen paintings, and are further illustrated with pictures of other plants found in that habitat. These single plant portraits are taken from the books illustrated by Barbara listed on page 77. Thus a little of the whole range of her plant portrait work is included and, by noticing the dates of original publication, the slight changes that occurred in her style and the improvements in their reproduction can be appreciated.

Woodlands

1   Pendunculate Oak, *Quercus robur*
2   Ash, *Fraxinus excelsior*
3   Wild Cherry, *Prunus avium*
4   Field Maple, *Acer campestre*
5   Broad Buckler-Fern, *Dryopteris dilatata*
6   Hazel, *Corylus avellana*
7   Hornbeam, *Carpinus betulus*
8   Enchanter's Nightshade, *Circaea lutetiana*
9   Wood Poa, *Poa nemoralis*
10   *Coriolus versicolor*
11   Ground Ivy, *Glechoma hederacea*
12   Bluebell, *Hyacynthoides nonscripta*
13   Honeysuckle, *Lonicera periclymenum*
14   Dog's Mercury, *Mercurialis perennis*
15   Yellow Archangel, *Lamiastrum galeobdolon*
16   Wood Melick, *Melica uniflora*
17   False Brome, *Brachypodium sylvaticum*
18   Ivy, *Hedera helix*
19   Lords-and-Ladies, *Arum maculatum*
20   *Hypogymnia physodes*
21   *Usnea florida*
22   *Parmelia sulcata*
23   *Parmelia caperata*
24   Wood Sage, *Teucrium scorodonia*
25   Common Dog-Violet, *Viola riviniana*
26   Lesser Celandine, *Ranunculus ficaria*
27   Wood Anemone, *Anemone nemorosa*
28   Primrose, *Primula vulgaris*
29   Wood-Sorrel, *Oxalis acetosella*
30   *Thuidium tamariscinum*
31   *Mnium hornum*
32   *Atrichum undulatum*
33   Sanicle, *Sanicula europaea*
34   Wood-Sedge, *Carex sylvatica*
35   Woolly Foot, *Collybia peronata*

Woodlands

This painting gives a general view of mixed woodland, with the plants that grow on the woodland floor shown in greater detail in the foreground. The main tree is the common oak (*Quercus robur*). There is another British species, *Quercus petraea*, sometimes called the durmast oak, that is commoner in the north, although the two of

Durmast oak, *Quercus petraea*

them sometimes grow together in southern England. *Quercus petraea* used to grow extensively just south of London, forming a great wood that has now been completely built over. The other large tree shown is ash (*Fraxinus excelsior*). Both oak and ask flower and come into leaf rather late, at about the beginning of May. The less common hornbeam (*Carpinus betulus*) flowers at about the same time.

Field maple (*Acer campestre*) flowers rather later, whereas the beautiful wild cherry (*Prunus avium*) is usually in full blossom for Easter. These two trees do not grow so tall as ash and oak, and form a separate, lower, layer in the woodland. Often there is also a shrub layer, with much lower growing shrubs such as hazel (*Corylus avellana*). Hazel flowers very early. Sometimes the catkins are shedding their pollen on New Year's Day, when the other trees and shrubs are leafless and inactive. The woodland floor is occupied by herbacious plants like dog's mercury (*Mercurialis perennis*) and bluebell (*Hyancinthoides nonscripta*). Then right at ground level are mosses and fungi, forming the lowest layer in this complex community.

In winter all plants other than the mosses and fungi are dormant, and it is the time of the year when most light can reach the woodland floor. For mosses it is a period of active growth. *Mnium hornum* is one of the commonest of woodland mosses, and often grows directly on the wood of tree roots, fallen logs and old stumps. Catherine's moss (*Atrichum undulatum*) is found on bare ground between the trees and along woodland rides. This handsome plant was associated with Catherine the Great of Russia by the eighteenth-century botanist Friedrich Ehrhart.

Although they do not require light for their growth, some fungi grow actively in winter. Clusters of sulphur tuft (*Hypholoma fasciculare*) appear on stumps and at the bases of trees. The bracket fungus *Coriolus versicolor* is very common on dead wood. Another kind of fungus, the earth ball (*Scleroderma aurantium*), grows directly on the ground, obtaining its nutriment from dead leaves and humus in the soil. Sometimes a small toadstool *Boletus parasiticus* will be found growing as a parasite on earth balls. All these fungi play an essential part in the community by breaking down fallen leaves and twigs into simpler materials which are available to promote the growth of other plants.

In spring, there is still plenty of light because

the various trees are not yet in full leaf, but the weather is warmer. Dog's mercury is wind pollinated and flowers early. With the appearance of the first bees begins a succession of familiar flowers such as common dog violet (*Viola riviniana*), celandine (*Ranunculus ficaria*) and primrose (*Primula vulgaris*). By May, when trees like ash and oak are coming into leaf, bluebells will be forming a carpet of colour in many woods. Although common in Britain, bluebells are not widespread on the continent of Europe and do not grow at all elsewhere, so they may be regarded as typically English flowers. They require open woodland, but do not survive if

Sulphur tuft, *Hypholoma fasciculare*

the wood is clear felled. The plants are also easily destroyed by trampling.

We know that some woodlands in Britain have been in existence for more than a thousand years in much the same state as they are today, yet they cannot be regarded as natural communities. From very early times they have been managed by methods known as coppicing and pollarding to produce poles for building purposes and smaller

wood for fuel. In coppicing, the trees were cut off completely at ground level. The shoots that grew up from the stools were left for five to ten years to grow into poles of a suitable size, and then all growth was cut down to the ground again. To prevent deer and cattle eating off the new growth, the woodland had to be securely fenced at least for a year or two after coppicing. Sometimes it was enclosed more or less permanently by surrounding the wood with a ditch and bank, and planting a hedge on top. Where secure fencing was not possible, pollarding was carried out instead. In pollarding, the trees were cut off at about shoulder height so that the new growth would be out of the reach of grazing animals.

These types of management have profound effects on the structure of the wood. Hazel and ash, for example, respond well to coppicing, producing excellent regrowth, and they are encouraged by the system. In some places sweet chestnut (*Castanea sativa*) was extensively planted, for it responds well too. In contrast, although some oaks were usually left to grow into timber trees, they were otherwise removed. Elm trees (*Ulmus procera*) were often planted in hedgerows and in time they would invade the woodland. The other woodland plants experienced a year or two immediately after coppicing when there was abundant sunlight and the atmosphere was fairly dry, followed by several years in which increasingly less light was available and the air became damper as the foliage of the trees created a denser canopy.

Coppicing is no longer an economic method of management in most parts of the country. Old coppice is converted to open woodland by thinning and sometimes by replanting. Increasingly, woods are being clear felled. As a result, deciduous woodland, the habitat of a great variety of insects, birds and other animals, is a diminishing resource. From the conservationist's point of view, management policies that permit selective rather than clear felling, and replanting with a mixture of species rather than creating plantations of a single species, are preferable to the conventional forestry practices that have been used so widely in the past.

Pine Forests

1 Scots Pine, *Pinus sylvestris*
2 Bilberry, *Vaccinium myrtillus*
3 Mat-Grass, *Nardus stricta*
4 Downy Birch, *Betula pubescens*
5 Hairy Wood-Rush, *Luzula pilosa*
6 Chickweed Wintergreen, *Trientalis europaea*
7 Heather, *Calluna vulgaris*
8 Lesser Twayblade, *Listera cordata*
9 Common Wintergreen, *Pyrola minor*
10 Twinflower, *Linnaea borealis*
11 Coralroot Orchid, *Corallorhiza trifida*
12 Heath Bedstraw, *Galium saxatile*
13 Creeping Lady's-Tresses, *Goodyera repens*
14 Rowan, *Sorbus aucuparia*
15 *Lactarius deliciosus*
16 *Lactarius rufus*
17 *Russula sardonia*
18 Tormentil, *Potentilla erecta*
19 Cowberry, *Vaccinium vitis-idaea*
20 *Plagiochila spinulosa*
21 *Rhytidiadelphus loreus*
22 *Barbilophozia floerkei*
23 *Ptilium crista-castrensis*
24 *Sphaerophorus globosus*
25 *Dicranum majus*
26 *Parmelia omphalodes*
27 *Hylocomium splendens*
28 *Plagiothecium undulatum*
29 Hard Fern, *Blechnum spicant*
30 *Bazzania tricrenata*
31 *Herberta adunca*

Pine Forests

The forest shown in this painting is a remnant of the great Caledonian forest that covered much of the Scottish Highlands from the retreat of the glaciers 10,000 years ago until relatively recent times. It consisted largely of pine and birch, although there was some oak on south-facing slopes and in valley bottoms.

One of the best-known remnants of Scottish pine forest is the Black Wood of Rannoch, which lies on the south shore of Loch Rannoch not far from Blair Atholl in Perthshire. It was visited by Barbara Nicholson when she was working on this painting. Like the rest of the Caledonian forest, the Black Wood has been exploited for timber over the past few hundred years. Large numbers of trees were cut down in the politically unsettled times of the seventeenth and eighteenth centuries. Further extensive felling took place during the Napoleonic Wars and also during the World Wars of this century. However, in between these destructive episodes the Wood has been able to recover to a considerable extent. Quite a number of its mature trees grew from seedlings that sprang up in the peaceful period following the Battle of Waterloo and there are many younger trees of various ages. There has been no planting—with the result that the trees of the Wood are a truly native population of pine (*Pinus sylvestris*).

Unlike trees in plantations, trees that have grown from naturally sown seedlings are not only of various ages but also show considerable variety in their growth form. In plantations the trees are all much alike because usually they are of a deliberately selected form taken from a known source. Some of the naturally occurring variation is shown here; in most specimens the branches are spreading more or less horizontally, but in some the branches grow upwards (form *ascensa*) and in others they drop (form *pendula*).

Pine tree seeds develop in female cones which

Cowberry, *Vaccinium vitis-idaea*

appear in spring at the tips of new shoots. The male cones that produce pollen develop at the base of other shoots. The female cone takes three years to ripen. As it matures it becomes brown and woody; eventually, before the cone itself falls to the ground, the scales open and the winged seeds are shed.

Birch plays a part in pine forests, and sometimes forms woods on its own. The downy birch (*Betula pubescens*) is the species generally found in the west and north of Scotland but tends to be replaced by silver birch (*Betula pendula*) in the Central and Eastern Highlands. Where a pine forest has been burnt, birches are frequently the first trees to recolonize and pines subsequently grow up amongst them. Rowan or mountain ash (*Sorbus aucuparia*), which is also frequently found with birch and pine, grows at higher altitudes than any other tree in Britain.

Ground cover on the acid soils of old pine forests is provided by three evergreen members of the heath family (Ericaceae). Heather (*Calluna vulgaris*) is frequently abundant. Bilberry (*Vaccinium myrtilis*) and cowberry (*Vaccinium vitis-idaea*), both of which have edible fruits, are tolerant of shade and often produce a very dense undergrowth with mosses such as *Hylocomium splendens*. In the more open forests of the Western Highlands, common mosses include *Plagiothecium undulatum* and *Ptilium crista-castrensis*. They may grow so densely that they impede natural regeneration of the trees.

Some fungi are characteristic of pine forests. The saffron milk cap (*Lactarius deliciosus*) has a faintly acrid taste and is certainly not delicious. *Russula sardonia* tastes very acrid, and should not be eaten.

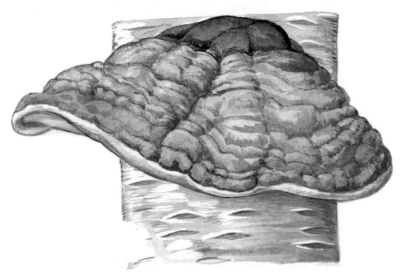

Bracket fungus, *Fomes fomentarius*

The bracket fungus *Fomes fomentarius* is found mainly in Scotland within the British Isles, where it only grows on birch trees.

A number of orchids grow in Scottish pine woods. Coral root (*Corallorhiza trifida*) is found in the Black Wood of Rannoch and elsewhere in the Eastern Highlands. This orchid, which requires rather damp conditions, is leafless and without any green colour. It is entirely dependent for its nutrition on a fungus living in its underground stems, which are fleshy and much branched and, rather fancifully, are thought to resemble coral. Creeping lady's tresses (*Goodyera repens*) and the lesser twayblade (*Listera cordata*) are found in drier conditions, the latter quite high on the hills and often partly hidden by heather and moss.

As attractive in its own way as an orchid is the common wintergreen (*Pyrola minor*). Although it is nowhere abundant, it is more common in Scotland than anywhere else in the British Isles. A related plant, one-flowered wintergreen (*Moneses uniflora*), is very rare within these islands and is found only in Scotland. Chickweed wintergreen (*Trientalis europaea*) is not closely related to the other two, but is very much an ancient pinewoods plant. Also restricted to native pinewoods, and found in Britain only in the east of Scotland, is the twinflower (*Linnaea borealis*). It was named

One-flowered wintergreen
Moneses uniflora

after the great Swedish naturalist Carl Linnaeus who, with rather exaggerated modesty, described it as 'a plant of Lapland, lowly, insignificant, disregarded, flowering but for a brief space—from Linnaeus, who resembles it.'

Mountains

1 Three-leaved Rush, *Juncus trifidus*
2 Mossy Saxifrage, *Saxifraga hypnoides*
3 Trailing Azalea, *Loiseleuria procumbens*
4 Alpine Saw-Wort, *Saussurea alpina*
5 *Amphidium mougeotii*
6 Dwarf Cudweed, *Gnaphalium supinum*
7 Holly Fern, *Polystichum lonchitis*
8 Fir Clubmoss, *Huperzia selago*
9 *Solorina crocea*
10 *Rhacomitrium lanuginosum*
11 Stiff Sedge, *Carex bigelowii*
12 Spiked Wood-Rush, *Luzula spicata*
13 *Polytrichum alpinum*
14 Alpine Mouse-Ear, *Cerastium alpinum*
15 Moss Campion, *Silene acaulis*
16 Roseroot, *Sedum rosea*
17 Rock Speedwell, *Veronica fruticans*
18 Net-leaved Willow, *Salix reticulata*
19 Sibbaldia, *Sibbaldia procumbens*
20 Dwarf Willow, *Salix herbacea*
21 *Andraea alpina*
22 *Anthelia julacea*
23 Alpine Clubmoss, *Diphasium alpinum*
24 Alpine Willowherb, *Epilobium anagallidifolium*
25 *Bryum pseudotriquetrum*
26 Mountain Avens, *Dryas octopetala*
27 *Plagiochila asplenioides*
28 Mat-Grass, *Nardus stricta*
29 Viviparous Fescue, *Festuca vivipara*
30 Alpine Meadow-Rue, *Thalictrum alpinum*
31 *Hylocomium splendens*
32 Green Spleenwort, *Asplenium viride*
33 Alpine Scurvy-Grass, *Cochlearia alpina*
34 *Rhizocarpon petraeum*
35 Purple Saxifrage, *Saxifraga oppositifolia*
36 Starry Saxifrage, *Saxifraga stellaris*
37 Mountain Sorrel, *Oxyria digyna*
38 *Peltigera venosa*
39 *Dicranum scoparium*
40 Crowberry, *Empetrum nigrum*
41 Alpine Lady's-Mantle, *Alchemilla alpina*
42 *Rhytidiadelphus loreus*
43 Alpine Bistort, *Polygonum viviparum*
44 Eyebright, *Euphrasia frigida*
45 Hoary Whitlow-Grass, *Draba incana*
46 *Pannaria hookeri*
47 Three-flowered Rush, *Juncus triglumis*
48 *Solorina saccata*

Mountains

Mountain areas provide harsh environments. Temperatures are frequently very low, although sunlight is often bright, and there is exposure to high winds all the year round. Several different plant communities are found on every mountain, depending upon particular conditions such as aspect, kind of soil, shelter, slope, rainfall and drainage.

The most mountainous region of Britain, the Scottish Highlands, provides the background to this painting. These mountains have a very long geological history. They are composed of many different rocks, some of which are among the most ancient known, and were formed in the remote past by vast earth movements and volcanic activity. Over an immense period of time they have been worn down by a ceaseless process of weathering, including severe periods of glaciation during the ice ages of the past two million years.

From a distance, the mountain peaks, which have only been free of permanent snow and ice for a few hundred thousand years, appear completely bare. However, even these exposed rocks have been colonized by plants, their surfaces being covered with crusts of firmly attached lichens. One of the most conspicuous of these is the map lichen (*Rhizocarpon geographicum*), which occurs widely at moderate and high altitudes on hard acid rock. Its name originated because the bright yellow-green surface layer develops on top of a very thin black under-layer, which shows as a black line at the margins and through cracks in the surface. When several plants grow together touching one another the black lines have the appearance of boundaries on a map. Other species of *Rhizocarpon* are found on basic, that is non-acid, rocks, for instance the common *Rhizocarpon umbilicatum* which forms neat, round, greyish-white patches with a pattern of cracks on the surface.

A few mosses are also capable of colonizing bare rock surfaces. On acid rocks, near the summit of mountains, *Andraea alpina* is quite common. It is a particularly interesting moss for, unlike any other, its spore capsule opens by four slits instead of the usual lid. Quite different species of moss grow on limestone, a common one being *Tortella tortuosa*, the leaves of which, when dry, have a conspicuous spiral twist.

A lichen of acid rock, *Rhizocarpon geographicum* (left), and two lichen of limestone rock, *Rhizocarpon umbilicatum* (right) and *Solenopsora candicans* (top).

Tortella tortuosa

Both mosses and lichens play a significant part in breaking down rock. As lichens and some mosses, such as *Andraea alpina*, spread slowly over the rock surface it gradually disintegrates and powders away. The fine particles thus formed mix with the larger particles produced by frost shattering to form the skeleton of a soil. On limestone the cushions of *Tortella tortuosa* slowly dissolve hollows in the rock, producing a characteristic weathered effect. The eventual death and decay of the pioneer plants produces humus, which accumulates amongst the skeleton soil and encourages the development of other species.

At the top left of the painting, a summit slope is shown covered with the larger debris of weathering. Between the boulders grows a carpet of the moss *Rhacomitrium lanuginosum*, which is often so extensive that it is referred to as '*Rhacomitrium* heath'. This sort of heath often includes plants such as the sedge *Carex bigelowii* and other flowering plants such as the dwarf cudweed (*Gnaphalium supinum*).

Saxifrages are able to grow over large stones and on small boulders provided that the rocks are sufficiently damp. *Saxifraga hypnoides*, often called 'Dovedale moss' because it is common on the Yorkshire limestone as well as in Scotland, is only found on basic rocks. Purple saxifrage (*Saxifraga oppositifolia*) grows on mountains up to altitudes of nearly 1,200 metres, and starry saxifrage (*Saxifraga stellaris*) is found in similar places and indeed even higher, for it is found on the summit of Ben Nevis, the highest mountain in the British Isles. Perhaps the most beautiful of all the flowering plants illustrated here is the mountain avens (*Dryas octopetala*). It grows on limestone and is found at all altitudes right down to sea level in the extreme north of Scotland. In late summer the handsome eight-petalled flowers are followed by clusters of plumed fruits.

A number of mountain grasses grow between rocks where pockets of soil have accumulated or on flatter areas where there is a fair depth of soil. In many places mat grass (*Nardus stricta*) covers large areas. It has very harsh foliage and is avoided by sheep. *Festuca vivipara* is a close relative of sheep's fescue (*Festuca ovina*) but differs from it in that the flowers are replaced by little plantlets, as can be seen in Barbara Nicholson's painting. These fall off eventually and grow into new plants. Something similar occurs in alpine bistort (*Polygonum viviparum*); some, though not all, of the flowers are replaced by little bulbils which grow directly into new plants when they become detached from the parent.

Among the spore plants to be found at high altitudes are the club mosses, diminutive modern descendants of a group that were great trees in coal measure times. In the fir club moss (*Huperzia selago*) the spore capsules are scattered amongst the foliage leaves, but in the alpine club moss (*Diphasium alpinum*) they are in green cones at the ends of the upright stems. Some ferns grow on limestone rocks, for instance green spleenwort (*Asplenium viride*) and holly fern (*Polystichum lonchitis*), which has evergreen glossy leaves. Net leaved willow (*Salix reticulata*), a very dwarf shrub with tiny flowers in catkins, also grows on limestone.

Dwarf willow (*Salix herbacea*) is found high up on practically every mountain in Scotland but the rock speedwell (*Veronica fruticosa*) is one of the specialities on Ben Lawers, though it is also found sparsely on other Scottish mountains. It has bright blue flowers, the largest of any British species of *Veronica*, and is the only one that is shrubby at the base.

Moors

1 Mat-Grass, *Nardus stricta*
2 Wavy Hair-Grass, *Deschampsia flexuosa*
3 Green-Ribbed Sedge, *Carex binervis*
4 Common Bent, *Agrostis capillaris*
5 Brown Bent, *Agrostis canina*
6 Heather, *Calluna vulgaris*
7 Purple Moor-Grass, *Molinia caerulea*
8 Sheep's Fescue, *Festuca ovina*
9 Bilberry, *Vaccinium myrtillus*
10 Bell Heather, *Erica cinerea*
11 Heath Rush, *Juncus squarrosus*
12 Northern Bilberry, *Vaccinium uliginosum*
13 *Campylopus pyriformis*
14 *Hypogymnia physodes*
15 Cowberry, *Vaccinium vitis-idaea*
16 Bearberry, *Arctostaphylos uva-ursi*
17 *Cladonia tenuis*
18 *Cladonia gracilis*
19 *Cladonia coccifera*
20 *Cladonia chlorophaea*
21 *Hygrophoropsis aurantiaca*
22 *Diplophyllum albicans*
23 *Ptilidium ciliare*
24 *Lophozia ventricosa*
25 *Dicranum scoparium*
26 *Laccaria laccata*
27 *Hypnum cupressiforme*
28 *Pleurozia schreberi*
29 Crowberry, *Empetrum nigrum*
30 *Clavaria argillacea*
31 *Coriscium viride*
32 *Omphalina hudsoniana*
33 *Polytrichum piliferum*
34 *Pohlia nutans*
35 Cross-leaved Heath, *Erica tetralix*
36 Tormentil, *Potentilla erecta*
37 Heath Bedstraw, *Galium saxatile*
38 *Campylopus atrovirens*
39 *Sphagnum compactum*
40 *Pleurozia purpurea*
41 Bog Myrtle, *Myrica gale*

Moors

Moorland, like bog and lowland heath, is a plant community of acid soils. Because of the abundance of heather (*Calluna vulgaris*) in the moorland plant community, it looks similar to heathland. Moorlands, however, develop in areas with a higher rainfall than heathland and are found widely in upland Britain. Clearly, with this painting, Barbara Nicholson intended to make us feel we are in Scotland, for nowhere else in Britain are there such extended views of hill country. Moorland, though, is found through a wide range of altitudes and latitudes—much of Yorkshire is or was moorland and there are also moors in south-west England.

Despite its wild character, moorland is for the greater part a managed habitat, used as a breeding and feeding ground for grouse and for grazing. Undisturbed moorland becomes covered with large heather bushes and only when individual plants die are gaps formed which can be colonized by other plants. Moorland management consists of controlled burning, carried out in small areas in regular rotation at intervals of from seven to twenty years, to increase the number of gaps. If the fire is not too fierce the woody roots of the burnt heather bushes send out fresh shoots which provide food for grouse. More intense burning of the top layer of peaty soil is disastrous because it destroys the roots and the dormant seeds in the soil. One of the hillsides in the background of the painting shows the typical patchwork pattern of moorland burnt on a regular rotation.

Often associated with heather on moorland are a number of related plants such as bell heather (*Erica cinerea*) and, in damper conditions, cross-leaved heath (*Erica tetralix*). Another, bilberry (*Vaccinium myrtillus*), is probably the commonest flowering plant of moorland after heather. Bilberries—known variously in different parts of Britain as whortleberries, huckleberries, whinberries and blaeberries—are pleasantly flavoured and are often used as fillings for pies and for making jams. Cranberries (*Vaccinium oxycoccos*) are used to make the traditional sauce served with pheasant, grouse and turkey. Bearberry (*Arctostaphylos uva-ursi*) and crowberry (*Empetrum nigrum*) produce fruit which are less palatable, but they do form part of the diet of moorland birds. Crowberry is a particularly tough and hardy plant. In some Yorkshire moors near towns where heather is unable to survive the destructive effects of trampling and pollution it has become the dominant plant.

Cranberry, *Vaccinium oxycoccos*

Grasses form an important element in the moorland community and provide grazing for deer and hare as well as sheep. Grasses such as sheep's fescue (*Festuca ovina*) and bent grasses (*Agrostis* species) are common grazing for sheep. However, moor grass (*Molinia coerulea*) and especially mat grass (*Nardus stricta*) are not liked by them; when the latter is abundant it is an indicator of overgrazing. Some grasses survive burning because the core of their dense tufts escapes the fire. Heath rush

Crowberry, *Empetrum nigrum*

greenish-grey, hollow and much-branched stems. Others, such as *Cladonia coccifera*, cover the ground with a layer of small green leaf-like scales and then produce cup-shaped structures on stalks. The spore-producing structures are bright red outgrowths formed on the rim of the cups. *Cladonia chlorophaea* and *Cladonia gracilis* are both brown-fruited species. The latter has long-stalked slender cups and sometimes smaller cups grow from the rims of the main ones. *Coriscium viride* also produces leaf-like scales but these are thicker and fleshier than those of *Cladonia* species, and are a brighter green. It only grows on damp peaty soil, and is associated with a small cream-coloured toadstool called *Omphalina hudsoniana*.

(*Juncus squarrosus*) is often associated with cross-leaved heath and sends up new growth from the roots after burning. Heather moorland often merges with grassland and frequent burning can encourage grasses at the expense of heather.

At lower altitudes it is probable that in the absence of burning and grazing much existing moorland would be colonized by juniper (*Juniperus communis*) and Scots pine (*Pinus sylvestris*) and ultimately change into woodland. This is prevented by traditional moorland management, but widespread successful planting of conifers on cleared moorland shows that this habitat can support trees. An area of conifer plantation can be seen in the background of the painting.

Non-flowering plants form a large part of the bottom layer of vegetation in moorland; mosses, liverworts and lichens colonize bare patches of soil. Most species of *Sphagnum*, a number of which are shown in the bog painting (pages 36–37), require water or at least very wet ground to grow in. *Sphagnum compactum* is an exception. It requires damp but not wet soil and seems to do particularly well after fire. Of the lichens, species of *Cladonia* are among the most conspicuous. Some of them, such as *Cladonia tenuis*, form quite large clumps of

Juniper, *Juniperus communis*
(Male cones, top right; shrub, bottom right)

31

Heaths

1   Gorse, *Ulex europaeus*
2   Bilberry, *Vaccinium myrtillus*
3   Dwarf Gorse, *Ulex minor*
4   Petty Whin, *Genista anglica*
5   Common Bent, *Agrostis capillaris*
6   Wavy Hair-Grass, *Deschampsia flexuosa*
7   Cross-leaved Heath, *Erica tetralix*
8   Bristle Bent, *Agrostis curtisii*
9   Broom, *Cytisus scoparius*
10   Bell Heather, *Erica cinerea*
11   Dorset Heath, *Erica ciliaris*
12   Cornish Heath, *Erica vagans*
13   Heather, *Calluna vulgaris*
14   Bracken, *Pteridium aquilinum*
15   Early Hair-Grass, *Aira praecox*
16   Heath Milkwort, *Polygala serpyllifolia*
17   Heath Dog-Violet, *Viola canina*
18   *Polytrichum juniperinum*
19   Heath Spotted-Orchid, *Dactylorhiza maculata*
20   Tormentil, *Potentilla erecta*
21   Heath Bedstraw, *Galium saxatile*
22   Dodder, *Cuscuta epithymum*
23   *Cladonia furcata*
24   *Diplophyllum albicans*
25   *Cladonia floerkeana*
26   *Baeomyces roseus*
27   *Cladonia uncialis*
28   *Cladonia portentosa*
29   Trailing St. John's-Wort, *Hypericum humifusum*
30   Wood Sage, *Teucrium scorodonia*
31   Heath Groundsel, *Senecio sylvaticus*
32   *Thelophora terrestris*
33   Horse-hair Fungus, *Marasmius androsaceus*
34   *Hypnum cupressiforme*
35   *Pleurozia schreberi*
36   Fairy Cakes, *Hebeloma crustuliniforme*

Heaths

Heathland plant communities develop on well-drained sandy soils in lowland Britain. These light soils are acid for several reasons. In the first place, the sands, sandstones and light gravels from which they are formed are deficient in basic minerals. As a result, the acid products produced by the decay of dead plant material are not fully neutralized. Furthermore, because of the rapid drainage, the minerals that are available tend to be washed out of the upper layers, which acquire a light grey colour. The leached materials are carried downwards to be redeposited as a dark reddish-brown layer in the subsoil. Sometimes sufficient accumulates to form an iron pan, a hard and impervious layer rich in iron oxide, which impedes drainage.

A common plant over much heathland is heather (*Calluna vulgaris*). It and other members of the heather family (Ericaceae) flourish on acid soils but grow only very poorly, if at all, on soils containing basic minerals such as chalk. The four British species of *Erica*, all of which are illustrated here, are commonly associated with heather but they would be found together only on heaths in Cornwall. The cross-leaved heath (*Erica tetralix*) is found in damp habitats. In Cornish heath (*Erica vagans*), the flowers point upwards instead of hanging down as in the others; it is found only in Cornwall. Dorset heath (*Erica ciliaris*) is rather more widespread and may be found from Dorset westwards to Land's End. Another related plant sometimes found on dry heathland is bilberry (*Vaccinium myrtillus*).

Larger shrubs such as broom (*Cytisus scoparius*) are common, especially where the ground has been disturbed. Gorse (*Ulex europaeus*) is related to broom but differs from it in being covered with sharp spines. Many attractive flowers grow amongst the shrubs. Trailing St. John's wort (*Hypericum humifusum*) is found throughout the British Isles. The closely related slender St. John's

wort (*Hypericum pulchrum*) is a very elegant plant with a similar distribution, although it is not so common. In damp places, the heath spotted orchid (*Dactylorhiza maculata* subspecies *ericetorum*) creates a striking effect.

Dodder (*Cuscuta epithymum*) is parasitic on gorse or heather. It is related to the bindweed but has no root system and only tiny leaves. The twining red stems, which are attached to the stem of the host, bear clusters of small waxy pink flowers.

Bare ground is colonized by lichens and mosses. The lichen *Baeomyces rufus* forms a greyish-green crust, and has brown spore-producing structures looking like tiny toadstools. Species of *Cladonia* are also common on bare ground. The spore-producing structures of *Cladonia floerkeana* are particularly conspicuous red outgrowths at the tips of the stalks. A number of lichens grow on heather twigs. A common and attractive one is *Hypogymnia physodes*.

A number of fungi are to be found on heaths in the autumn; *Thelophora terrestris* forms dark-brown clusters of fringed fronds, often flushed with violet tints. The toadstool *Hebeloma crustuliniforme* has a distinct smell of radishes; however, it should not be eaten. The charcoal toadstool (*Pholiota carbonaria*) is associated with pine trees and other conifers, and appears especially on burnt ground.

The moss *Polytrichum juniperinum* has spore cases that develop under hat-like coverings. As the stalks

Slender St. John's wort, *Hypericum pulchrum*

grow longer, the 'hats' fall off and the spore cases take up an almost horizontal position. The male organs develop in shallow cups formed from rosettes of leaves at the tips of the stalks and look almost like little flowers. *Polytrichum* species have unbranched stems that grow upright, and when they are not producing sexual organs and spore cases, plants growing in a clump look rather like a miniature forest of pine trees. *Pleurozium schreberi* is a moss with a quite different growth form. It does not grow upright, and the much-branched stems form a weft over the surface of the ground and amongst the stems of other plants. It can be distinguished from somewhat similar mosses that may be growing with it, such as *Hypnum cupressiforme*, because the stems are bright red colour. *Hypnum cupressiforme* is a very common moss that often forms dense, rather dark green mats on the ground and on tree trunks and walls, and it is very variable. The variety that grows on heaths and is shown in the painting is more slender, a paler green colour, and forms wefts instead of mats. The liverwort *Diplophyllum albicans* forms little turfs of crowded upright shoots that grow from creeping pale-green stems.

Heathland grasses are equally attractive on a larger scale. The flower clusters of wavy hair grass

Baeomyces rufus

(*Deschampsia flexuosa*), which appear in summer, have fine, thread-like, wavy branches, and can be used in floral decorations to good effect. Early hair grass (*Aira praecox*) is somewhat similar, but it is a much smaller plant and completes its life cycle in spring. Both these species are widespread throughout Britain. Bristle bent (*Agrostis setacea*) is a grass of the south and west of England and Wales that often forms swards of soft, springy turf.

In the past fifty years the area of heathland in southern England has been drastically reduced. Much of it has been taken over for housing, particularly near the sea. Further inland, extensive tracts have been used for motorways; road alignments that cross heath avoid highly developed and expensive agricultural land. Although the light soils of heaths do not produce high yields from farm crops, they are suitable for forestry, and large areas have been taken over for plantations, particularly of conifers. The remaining inland heaths are quite small and fragmented. At the seaside, they are increasingly used for holiday camps and caravan parks. The areas of heath that do survive are very popular with ramblers, picnickers and amateur naturalists; but unfortunately the heathland habitat is very easily eroded by trampling and is highly susceptible to destruction by fire. Heath plant communities have become the most severely threatened of all in Britain, and it is important that those heaths that remain should be conserved as fully as possible.

Charcoal toadstool, *Pholiota carbonaria*

Bogs

1 Hare's-Tail Cottongrass, *Eriophorum vaginatum*
2 Heather, *Calluna vulgaris*
3 Cross-leaved Heath, *Erica tetralix*
4 Purple Moor-Grass, *Molinia caerulea*
5 Great Sundew, *Drosera anglica*
6 Oblong-leaved Sundew, *Drosera intermedia*
7 *Sphagnum subnitens*
8 Round-leaved Sundew, *Drosera rotundifolia*
9 *Sphagnum capillifolium*
10 Bog Myrtle, *Myrica gale*
11 Bog Asphodel, *Narthecium ossifragum*
12 *Sphagnum auriculatum*
13 *Aulacomnium palustre*
14 *Sphagnum magellanicum*
15 Tormentil, *Potentilla erecta*
16 *Polytrichum alpestre*
17 *Sphagnum recurvum*
18 *Cladonia arbuscula*
19 *Cladonia uncialis*
20 *Leucobryum glaucum*
21 Cranberry, *Vaccinium oxycoccos*
22 *Odontoschisma sphagni*
23 *Sphagnum papillosum*
24 *Lepidozia setacea*
25 *Mylia anomala*
26 *Calypogeia trichomanis*
27 *Sphagnum tenellum*
28 White Beak-Sedge, *Rhynchospora alba*
29 Bog Rosemary, *Andromeda polifolia*
30 Common Cottongrass, *Eriophorum angustifolium*
31 *Mylia taylori*
32 Common Butterwort, *Pinguicula vulgaris*
33 *Sphagnum cuspidatum*
34 Bog Pondweed, *Potamogeton polygonifolius*
35 Bog Sedge, *Carex limosa*
36 Deergrass, *Scirpus cespitosus*
37 *Campylopus paradoxus*
38 Lesser Bladderwort, *Utricularia minor*

Bogs

The ancient, hard rocks of the north and west of the British Isles are said to be acid because they lack minerals that will readily dissolve in rain water. They weather slowly to produce thin soils that are also acid. As the rainfall is quite high, and in many places the drainage is poor, these soils become permanently waterlogged. Most plants cannot thrive in such wet, acid conditions, except for a group known as 'bog mosses' (*Sphagnum* species). These grow luxuriantly to form dense turfs, and make up a community known as blanket bog. Over long periods of time the sphagnum stems grow taller and taller, and the older, lower parts die. Because the acid conditions discourage the development of most of the organisms responsible for decay, mainly bacteria and fungi, breakdown of the dead stems is very slow, but eventually they decay sufficiently to form deposits of peat.

In places where there is acid marshy ground around springs and beside streams, valley bog may develop, even when the rainfall is not very high. There is always a slow current of water and a slightly higher mineral content than in bogs where the water is completely stagnant. This produces a somewhat different plant community from that found in blanket bog.

Another kind of bog is called raised bog. This may develop in marshy places that originally were not acid. The marsh becomes choked with fen vegetation and the products of decayed dead plant material causes acid conditions in the upper layers. The growth of sphagnum species that can tolerate the rather drier conditions at the surface produces a bog shaped like a shallow dome.

When making her preliminary sketches for this painting, Barbara Nicholson visited Borth Bog, a National Nature Reserve in mid Wales. Her painting shows a valley bog in the middle distance and a raised bog in the background. The hummock in the foreground is of a kind found in all types of bog.

Formation of hummocks occurs because the vigorous growth of a species can change the conditions of the environment sufficiently to make them suitable for another species. *Sphagnum cuspidatum*, for instance, can colonize more or less open water in shallow pools. When it has formed a dense mat, *Sphagnum subnitens* is able to grow on top of it, forming a clump, and *Sphagnum capillifolium* can grow on top of that. In the slightly drier conditions that result, lichens, such as *Cladonia arbuscula*, and perhaps the fungus *Omphalina sphagnicola*, may appear. Some species of sphagnum are very precise in their requirements. For example, *Sphagnum pulchrum* is only found in valley bogs and in wet hollows in raised bogs. *Sphagnum subnitens*, on the other hand, is the commonest species in both blanket and raised bog.

In summer, large areas of bog throughout the British Isles are yellow with the flowers of bog asphodel (*Narthecium ossifragum*). In spring, the most conspicuous plants are the cottongrasses, particularly the white blobs of *Eriophorum vaginatum*. The catkins of bog myrtle (*Myrica gale*) appear at the same time; the strongly aromatic leaves appear later. The beautiful

Sphagnum cuspidatum

Bog rosemary, *Andromeda polifolia*

bog rosemary (*Andromeda polifolia*), has flowers that are flushed a delicate pink colour. Linnaeus borrowed the name Andromeda from the Greek nymph who was chained naked to a rock as a sacrifice to a sea monster, and who he thought would have blushed similarly when the hero Perseus came to rescue her. Linnaeus made a sketch in his Lapland notebook of Andromeda (the nymph) and Andromeda (the flower) and labelled them 'fiction and truth'.

The cottongrasses (*Eriophorum* species) are not the only members of the sedge family (Cyperaceae) that flourish in bogs. A typical example of a sedge that is shown in the painting is white beak sedge (*Rhynchospora alba*). It is a slender and attractive grass-like plant which forms tufts. The flowers grow in dense pale straw-coloured clusters. It can easily be distinguished from a grass because the stems are triangular in cross-section. Bog sedge (*Carex limosa*)—which also has triangular stems— does not grow in tufts, but is a far-spreading plant of very wet places. The male flowers grow in upright spikes, and the clusters of female flowers hang downwards. Deergrass (*Scirpus cespitosus*) is another member of the sedge family, but the stems are round in cross-section and when the plant is not in flower can be mistaken at first glance for the leaves of a grass. It often grows in extensive patches in blanket bogs. It may grow in very wet places too, and can be grown successfully as an aquarium plant.

A number of bog plants make good the deficiency of mineral nutrients by trapping and digesting small insects and other tiny animals. Butterwort (*Pinguicula vulgaris*) does this in the simplest way. The upper surface of the leaf has the appearance of having been very thinly spread with butter. Very small insects become trapped by the stickiness; larger ones can struggle away without much difficulty. The leaf produces protein-digesting enzymes, and the insects that do not escape are killed by them and gradually digested. The chopped leaves of butterwort have been used to turn warm milk into the sweet dessert, junket.

Sundew (*Drosera* species) have a more elaborate mechanism for trapping and digesting insects. The leaves are covered with tentacles that have round glands at their tips. When an insect becomes stuck on one or two of the tentacles, the others bend over so as to make contact. The process is not very rapid but in a few moments most of the tentacles on the leaf will be stuck to the prey. The leaf itself then rolls up with the insect inside and digestive juices are poured on to it from the glands. All three British species are shown. The commonest British sundew is *Drosera rotundifolia*, which is found in very wet as well as somewhat drier conditions. *Drosera anglica*, which is a plant of wet places, is commoner in Scotland than in England. The third British species, *Drosera intermedia*, grows mainly in the west of the British Isles and prefers drier places such as the tops of hummocks.

Bladderwort (*Utricularia minor*) grows entirely submerged in water except for the flower stalk, which carries the yellow flowers well above the surface. The leaves are finely divided and on some of them are the bladders in which tiny animals are trapped and digested.

39

Fens

1 Alder Buckthorn, *Frangula alnus*
2 Grey Willow, *Salix cinerea*
3 Guelder-Rose, *Viburnum opulus*
4 Marsh Fern, *Thelypteris palustris*
5 Yellow Iris, *Iris pseudacorus*
6 Marsh Thistle, *Cirsium palustre*
7 Reed Grass, *Phalaris arundinacea*
8 Buckthorn, *Rhamnus catharticus*
9 Purple Small-Reed, *Calamagrostis canescens*
10 Downy Birch, *Betula pubescens*
11 Blunt-flowered Rush, *Juncus subnodulosus*
12 Alder, *Alnus glutinosa*
13 Black Currant, *Ribes nigrum*
14 Common Reed, *Phragmites australis*
15 Lesser Reedmace, *Typha angustifolia*
16 Common Club-Rush, *Scirpus lacustris*
17 Greater Tussock-Sedge, *Carex paniculata*
18 Lesser Pond-Sedge, *Carex acutiformis*
19 Great Fen-Sedge, *Cladium mariscus*
20 Common Marsh-Bedstraw, *Galium palustre*
21 Greater Birdsfoot-Trefoil, *Lotus uliginosus*
22 Marsh Pennywort, *Hydrocotyle vulgaris*
23 *Campylium stellatum*
24 *Calliergon cordifolium*
25 Greater Spearwort, *Ranunculus lingua*
26 Yellow Loosestrife, *Lysimachia vulgaris*
27 Meadowsweet, *Filipendula ulmaria*
28 Tufted-Sedge, *Carex elata*

Fens

Fens develop on soils rich in plant nutrients where the water table is permanently high. When drained, fen becomes very productive agricultural land; in its waterlogged condition it supports a highly characteristic community of plants.

There is still quite a considerable extent of fen in north-east Ireland, but in England few remnants have survived the efficient drainage of lowland areas. Much of East Anglia was once fenland—now only patches of the original fen remain. One of the best known is Wicken Fen in Cambridgeshire, a property of the National Trust, which Barbara Nicholson visited when working on this painting. Here almost all the stages in the development of fen can be seen. Wicken Fen's status as a nature reserve has helped to ensure the survival there of plant communities that are becoming increasingly uncommon.

Drainage of the East Anglian fens began in the seventeenth century and has been maintained and extended ever since. A consequence of drainage has been a lowering of the soil level caused by shrinkage and decomposition of the peat. A complementary pumping system has therefore been necessary to raise the drainage water to the level of the water in the drainage channels. Before the advent of steam engines and later diesel and electric motors to drive the pumps, wind power was used, and a wind pump is shown in the picture.

The open water of the East Anglian fens is found in the drainage channels known locally as lodes, and also in straight-sided pools. These were formed by cutting peat for fuel and in some places by digging through the peat to obtain the clay from below for brick-making. Many aquatic plants are found in this open water. Conspicuous among them are white water lily (*Nymphaea alba*) which has round floating leaves and showy flowers and, in shallower water, tubular water dropwort (*Oenanthe fistulosa*) which has finely divided leaves and massed small flowers. A lode is shown in the distance in the painting, with an alder tree (*Alnus glutinosa*) growing on the bank, and a brick pool in the foreground, with meadowsweet and yellow loosestrife growing beside it.

The remains of dead plants do not decay quickly under water and accumulate as peat. Reeds (*Phragmites australis*) colonize the shallower water, crowding out other plants and forming extensive beds. Peat continues to accumulate, the water becomes shallower still, and great fen sedge (*Cladium mariscus*) takes over from the reeds. In the past, reed and sedge were cut to provide thatching material for

White water lily, *Nymphaea alba*

42

roofing and to some extent the industry still continues. In 'Norfolk thatch' reed is used as the main cover and sedge, which is longer-lasting but more expensive, is used for ridging. The painting shows reeds being cut and stacked on a barge.

The removal of invading bushes on old sedge fields, as is done in parts of Wicken, maintains a community known as 'mixed fen'. It includes many attractive flowers such as meadowsweet (*Filipendula ulmaria*), yellow loosestrife (*Lysimachia vulgaris*) and the rather less common greater spearwort (*Ranunculus lingua*). At Wicken, broad pathways known as droves are mown regularly and have been so in some cases for hundreds of years. Among the plants found in these conditions are tormentil (*Potentilla erecta*) and meadow thistle

Tormentil, *Potentilla erecta*

(*Cirsium dissectum*). The latter differs from the much commoner marsh thistle (*Cirsium palustre*) in having solitary rather than clustered heads and the stem does not have spiny wings.

When fen is not cut and kept free from bushes it gradually develops into fen scrub or carr. Alder buckthorn (*Frangula alnus*) is generally the pioneer shrub colonist. Once established it can soon form

almost impenetrable thickets. As the carr matures, alder buckthorn persists but other shrubs also colonize. Buckthorn (*Rhamnus catharticus*) is a common plant in mature fen carr. Unlike alder buckthorn, it has spiny tips to its branches, toothed leaves and wood that when cut stains bright orange; the cut wood of alder buckthorn stains bright lemon yellow. Other carr shrubs include the guelder rose (*Viburnum opulus*) and the grey willow (*Salix cinerea*). The subspecies illustrated here, *cinerea*, is relatively uncommon except in the East Anglian fens. The much more widespread subspecies *oleifolia* always has rusty brown hairs on the underside of the leaves. In carr the soil is still water-logged but many of the attractive flowers of the open fen do not survive because there is insufficient light. Shade tolerant plants flourish, though, including the marsh fern (*Thelypteris palustris*) which has far-creeping underground stems and spreads freely among the bushes.

Ultimately fen carr will give way to woodland, the final stage in the succession being shown at the top right of Barbara Nicholson's painting. Saplings of trees such as downy birch (*Betula pubescens*) and alder (*Alnus glutinosa*) become established in the carr and later such species as ash (*Fraxinus excelsior*) and oak (*Quercus robur*). However, human intervention has ensured that the succession of plant communities from open water to dry land rarely culminates in oak-ash woodland.

Meadow thistle, *Cirsium dissectum*

43

Freshwater

1 Greater Pond-Sedge, *Carex riparia*
2 Reed Grass, *Phalaris arundinacea*
3 Yellow Iris, *Iris pseudacorus*
4 Reedmace, *Typha latifolia*
5 Alder, *Alnus glutinosa*
6 White Water-Lily, *Nymphaea alba*
7 Crack Willow, *Salix fragilis*
8 Great Willowherb, *Epilobium hirsutum*
9 Purple Loosestrife, *Lythrum salicaria*
10 Yellow Water-Lily, *Nuphar lutea*
11 Common Club-Rush, *Scirpus lacustris*
12 Arrowhead, *Sagittaria sagittifolia*
13 Water Plantain, *Alisma plantago-aquatica*
14 Flowering-Rush, *Butomus umbellatus*
15 Common Reed, *Phragmites australis*
16 Marsh Cinquefoil, *Potentilla palustris*
17 Broad-leaved Pondweed, *Potamogeton natans*
18 *Brachythecium rivulare*
19 Marsh-Marigold, *Caltha palustris*
20 Water Horsetail, *Equisetum fluviatile*
21 Bogbean, *Menyanthes trifoliata*
22 Common Duckweed, *Lemna minor*
23 Water Forget-Me-Not, *Myosotis scorpioides*
24 Branched Bur-Reed, *Sparganium erectum*
25 Amphibious Bistort, *Polygonum amphibium*
26 *Fontinalis antipyretica*
27 Ivy-leaved Duckweed, *Lemna trisulca*
28 Water Mint, *Mentha aquatica*

Freshwater

The wetland habitats of the British Isles have been diminishing for a very long time. In southern England, only in relatively unimproved areas like the New Forest is it possible to get some idea of how wet much of the land must once have been over most of the country. There, both in the woods and in the open spaces, the water table in the soil is very near the surface for much of the year. Almost everywhere else it has fallen drastically due to artificial drainage. In East Anglian fens, drainage was begun in the seventeenth century, and in the wetlands associated with big river systems, such as the Mersey and the Dee in the north-east and the Thames in the south, it was well under way in the eighteenth century. The construction of canals, ornamental lakes and reservoirs does not make up for the loss of natural aquatic habitats, though some of the plants illustrated can be found in them.

Wetlands frequently silt up, and eventually trees establish themselves on the wet ground. Typical waterside trees are alder (*Alnus glutinosa*) and the willows such as the crack willow (*Salix fragilis*), so-called because its twigs are very easily broken at the junctions. Amongst the most conspicuous herbaceous plants by the waterside are the great willowherb (*Epilobium hirsutum*) and purple loosestrife (*Lythrum salicaria*). Others grow in shallow water, including the spring-flowering marsh marigold (*Caltha palustris*), yellow iris (*Iris pseudacorus*), water forget-me-not (*Myosotis scorpioides*) and marsh cinquefoil (*Potentilla palustris*). Two introduced plants, monkey flower (*Mimulus guttatus*) from North America and Himalayan balsam (*Impatiens glandulifera*), have become widespread in similar habitats throughout the country.

The dominant plants fringing open water and growing in the shallows are generally reeds, rushes and sedges. Among the most frequently found are common reed (*Phragmites australis*), reed grass (*Phalaris arundinacea*), reedmace (*Typha latifolia*),

pond sedge (*Carex riparia*) and club-rush (*Scirpus lacustris*), the latter being found in deeper water. Flowering rush (*Butomus umbellatus*), another plant of the water's margin, is called 'Pride of the Thames' in Dorset where Barbara Nicholson lived. In similar places, bogbean (*Menyanthes trifoliatea*) may be found. The leaves resemble those of a broad bean and the flowers are amongst the most beautiful of all wild flowers.

Water forget-me-not, *Myosotis scorpioides*

Hemp agrimony, *Eupatorium cannabinum*

Water horsetail (*Equisetum fluviatile*), is a spore-bearing plant whose ancestors were great trees in coal measure times. The spores are produced in cones at the tops of the stems as shown here. It sometimes forms quite large colonies in shallow water. Mare's tail (*Hippuris vulgaris*) is superficially similar but it has tiny flowers arranged in a ring just above the leaves. It is usually found growing in fairly fast-moving water.

Some plants can grow in water or on comparatively dry land. Amphibious bistort (*Polygonum amphibium*), for instance, has aquatic and terrestrial forms that are quite distinct. The former has floating stems and leaves and is hairless, while the latter is much more compact and is slightly hairy. Hemp agrimony (*Eupatorium cannabinum*) grows commonly on river banks, but it is found also in fens and moist woodlands. It was the basis of old herbal remedies for coughs and other illnesses. The famous herbalist, Nicholas Culpeper, writing in 1669, thought it to be 'inferior to but few herbs that grow'.

Some freshwater plants grow always entirely submerged, for instance the very common and well-known introduced plant, Canadian pond weed (*Elodea canadensis*), which is not illustrated here. The only plant of this kind that is shown in the painting is the moss *Fontinalis antipyretica*. It is easily the largest moss that grows in Britain, for it sometimes forms dense bunches as long as a man's arm. Linnaeus gave it the name 'antipyretica' because it was used as an almost completely fireproof insulating material in the walls of houses in Lapland. Many pondweeds (*Potamogeton* species) grow completely submerged. The broad-leaved pondweed (*Potamogeton natans*), however, has leaves that float on the surface of the water as well as submerged ones. The floating leaves have a tough texture and a shiny upper surface from which the water runs off easily if the leaf is accidentally submerged.

The two British species of water lily have similar floating leaves although, like the pondweeds, they are firmly rooted on the bottom. The leathery, shining surfaced leaves of white water lily (*Nymphaea alba*) are round, whereas those of yellow water lily (*Nuphar lutea*) are an oval shape. The latter also has submerged leaves, which are thin and translucent, and look something like lettuce. The flask-shaped fruits smell of alcohol as they ripen, which accounts for the common name of brandy bottle in some parts of the British Isles. Arrowhead (*Sagittaria sagittifolia*) is another plant of fairly deep water that has two kinds of leaves. Those that are exposed have the characteristic shape shown in the illustration but the submerged leaves are ribbon-shaped. Ribbon leaves, like those that are very small or finely divided, are adaptations to flowing water as they offer less resistance to the current.

Almost everywhere in Britain the plant populations of water habitats are in decline, due to pollution of various kinds. It is not only waste from industry and sewage works that may pollute rivers. Modern agricultural methods involve the use of herbicides and insecticides and the application of artificial fertilizers. The run-off of these is now a major threat to freshwater plant communities.

Salt Marshes

1 Sea-Purslane, *Halimione portulacoides*
2 Beaked Tasselweed, *Ruppia maritima*
3 Annual Sea-Blite, *Suaeda maritima*
4 Sea Aster, *Aster tripolium*
5 Sea Aster, *Aster tripolium* var. *discoideus*
6 Narrow-leaved Eelgrass, *Zostera angustifolia*
7 Common Saltmarsh-Grass, *Puccinellia maritima*
8 Common Sea-Lavender, *Limonium vulgare*
9 Sea Arrowgrass, *Triglochin maritima*
10 Townsend's Cord-Grass, *Spartina townsendii*
11 Shrubby Sea-Blite, *Suaeda fruticosa*
12 Greater Sea-Spurrey, *Spergularia media*
13 Sea-Heath, *Frankenia laevis*
14 Lesser Sea-Spurrey, *Spergularia marina*
15 Saltmarsh Rush, *Juncus gerardii*
16 Sea Wormwood, *Artemisia maritima*
17 Hard-Grass, *Parapholis strigosa*
18 Sea Rush, *Juncus maritimus*
19 Creeping Bent, *Agrostis stolonifera*
20 *Bostrychia scorpioides*
21 Glasswort, *Salicornia europaea*
22 Bladder Wrack, *Fucus vesiculosus*
23 Channelled Wrack, *Pelvetia canaliculata*
24 *Enteromorpha torta*
25 Perennial Glasswort, *Sarcocornia perennis*
26 Sea Plantain, *Plantago maritima*
27 *Enteromorpha intestinalis*
28 English Scurvy-Grass, *Cochlearia anglica*
29 Red Fescue, *Festuca rubra*
30 Buck's-Horn Plantain, *Plantago coronopus*
31 Sea-Milkwort, *Glaux maritima*

Salt Marshes

Salt marshes develop in estuaries where rivers deposit their load of silt and clay particles on reaching the sea. The mudbanks formed by water-borne deposits are shifted continually, mud being carried some distance upstream by the incoming tide and being scoured out again by the river outflow when the tide ebbs. This movement of mud would be very much more extensive were it not for the stabilizing effect of the salt marsh plants that grow on the areas exposed at low tide.

The impressive estuary and extensive stretches of salt marsh of the Thames, Britain's longest river, provide the background of this painting. Industrial, agricultural and housing developments have greatly altered the appearance of the dry land, but a considerable area of the salt marsh survives despite the modern flotsam of seemingly indestructible nylon rope and plastic containers.

An early colonizer of the mud in some places is eelgrass (*Zostera angustifolia*). This is a remarkable plant which lives on the firm mudbanks just above the low-tide level. Flowering, pollination, seed-set and seed dispersal all take place under water. The pollen does not consist of grains but is in the form of minute threads, which are carried from flower to flower by water currents. The tiny seeds are dispersed in the same way.

The usual pioneer colonizers of salt marsh are called glassworts (*Salicornia* species). There are about five species to be found in the Thames estuary, of which the most common is *Salicornia europaea*. All of them have tenacious roots and fleshy stems. The leaves are fleshy too and grow in opposite pairs joined at the edges and to the stems. The tops of the plants can be boiled and eaten in the same way as asparagus but are decidedly inferior in taste.

A number of other salt marsh plants are members of the same family as *Salicornia*, for instance sea purslane (*Halimione portulacoides*) and sea blite

Suaeda species). The sea beet (*Beta vulgaris* subspecies *maritima*) is a wild relation of the cultivated beetroot and sugar beet.

As soon as they start to grow in the spring the *Salicornia* plants slow down the tidal movements of the mud and, as their stems lengthen, fresh mud collects round them. The shifting mudbanks become more stable and their height increases too. In many places, however, their place is taken by cord grass (*Spartina anglica*). It is a very vigorous grass, a hybrid between a European and an American species, which first appeared in Southampton Water at the beginning of this century. It has spread widely and is capable not only of the primary colonization of the bare mud but of forming a continuous bed from low to high water.

Typical plants of the middle marsh are sea blite (*Suaeda maritima*) and sea aster (*Aster tripolium*). The latter, a relative of the Michaelmas daisy found in gardens, has quite showy purplish-blue flower heads.

Sea beet, *Beta vulgaris*

White flowers are not uncommon, however, and a variety *discoideus* with heads that lack ray florets and have only the yellow disc ones is shown here with the ordinary form. Even more conspicuous is golden samphire (*Inula crithmoides*) which has bright yellow flowers. This is a plant which grows predominantly on the south and west coasts but also appears on salt marshes in Essex and Kent. The plants of this middle zone are inundated by the sea twice a day but only rarely are they covered to a depth of more than a few centimetres.

These plants further consolidate the marsh and raise its level. Plants of the older, stable marsh include sea lavender (*Limonium vulgare*), which has purplish-blue 'everlasting' flowers, sea plantain (*Plantago maritima*) and sea purslane (*Halimione portulacoides*). Sea purslane is a shrubby perennial which grows especially beside channels and pools. In southern and eastern England sea blite (*Suaeda fruticosa*) and perennial glasswort (*Sarcocornia perennis*) also grow on the high marsh. High up on the beach red fescue grass (*Festuca rubra*) often forms a continuous sward where it is rarely reached by the sea.

Behind the high marsh there is frequently a wet area, sometimes merging with freshwater reed swamp, in which the rushes *Juncus maritimus* and

Golden samphire
Inula crithmoides

Juncus gerardii may be conspicuous. Among other plants found in this habitat are the sea-milkwort (*Glaux maritima*), sea wormwood (*Artemisia maritima*) and creeping bent (*Agrostis stolonifera*).

Some spore plants as well as flowering plants are found on salt marshes. The brown seaweeds (brown algae) known as 'wracks' are very common on rocky shores and grow everywhere that they can secure a hold. Two in particular, channelled wrack (*Pelvetia canaliculata*) and bladder wrack (*Fucus vesiculosus*), have small forms that are often found among other plants at middle levels in salt marshes. *Bostrychia scorpioides* is a small red alga that is found only in this habitat. These algae are unusual in living practically all the time out of the water. The green algae *Enteromorpha intestinalis* and *Enteromorpha torta* grow in brackish water and are found in the upper parts of salt marsh creeks. They often penetrate into the drainage dykes in the saltings above the marsh and they are indicators of a fairly high salt content in the water, perhaps resulting from incursions by the sea.

A salt marsh is at its best in the autumn. It is the time of year at which it has reached its fullest development, for many of the plants flower late. When winter comes, the annual plants die, of course, and the harsh weather sweeps away the vegetation of the lower levels of the marsh, replacing it with bare mud. In the spring, the annual species of *Salicornia* reappear, and the marsh builds up again. Inland, it would take, say, five years for a piece of bare ground to become grassland, ten years to become a scrub thicket, and a hundred years or more to become a recognizable woodland. On a salt marsh all the stages can be seen at once, from bare mud to red fescue grassland, at different levels on the beach. In between can be seen the various plant communities shown in this painting, including plants that are beautiful, such as sea heath (*Frankenia laevis*) and sea lavender (*Limonium vulgare*), and others that are particularly interesting, such as the members of the family Chenopodiaceae (glassworts, sea blites, sea purslane) that show specialized adaptations to an unusual and rather harsh environment.

Sand Dunes

1 Sand Sedge, *Carex arenaria*
2 Sea-Holly, *Eryngium maritimum*
3 Curled Dock, *Rumex crispus*
4 Marram, *Ammophila arenaria*
5 Lyme-Grass, *Leymus arenarius*
6 Sea Rocket, *Cakile maritima*
7 Sea Sandwort, *Honkenya peploides*
8 Prickly Saltwort, *Salsola kali*
9 Yellow Horned-Poppy, *Glaucium flavum*
10 Sea Spurge, *Euphorbia paralias*
11 Burnet Rose, *Rosa pimpinellifolia*
12 Lady's Bedstraw, *Galium verum*
13 Sea Buckthorn, *Hippophae rhamnoides*
14 Common Ragwort, *Senecio jacobaea*
15 Carline Thistle, *Carlina vulgaris*
16 Viper's-Bugloss, *Echium vulgare*
17 *Tortula ruraliformis*
18 *Peltigera canina*
19 Silverweed, *Potentilla anserina*
20 Common Birdsfoot-Trefoil, *Lotus corniculatus*
21 Kidney-Vetch, *Anthyllis vulneraria*
22 Haresfoot Clover, *Trifolium arvense*
23 Sand Couch, *Elymus farctus*
24 Portland Spurge, *Euphorbia portlandica*
25 Sea Bindweed, *Calystegia soldanella*
26 Sand Catstail, *Phleum arenarium*
27 Wild Pansy, *Viola tricolor* subsp. *curtisii*
28 Catsear, *Hypochoeris radicata*
29 Biting Stonecrop, *Sedum acre*
30 Wild Thyme, *Thymus drucei*
31 Common Restharrow, *Ononis repens*
32 Red Fescue, *Festuca rubra*
33 Common Centaury, *Centaurium erythraea*
34 Sea Fern-Grass, *Desmazera marina*
35 Common Storksbill, *Erodium cicutarium*
36 *Cladonia chlorophaea*

Sand Dunes

Sandy beaches are common all round the coasts of the British Isles, but on most of them the sand is inundated by the sea at every tide. Where the land is flat, ridges of sand are thrown up above the high water mark. If the prevailing wind comes from the right quarter, sand is blown inland and the ridges are built up into high dunes. Suitable conditions for the formation of dunes exist, for example, on the north coasts of Cornwall and Devon, where the beaches run at right angles to the fetch of the sea and the prevailing westerly winds. Among the most extensive dune systems in Britain are those at the estuaries of rivers as at Braunton Burrows, near Barnstaple in north Devon, and at Studland, near Poole in Dorset. Both of these localities were visited by Barbara Nicholson when she was working on this painting.

An early colonizer of newly formed dunes is sand sedge (*Carex arenaria*). It has tough wiry underground stems that run along in straight lines beneath the sand throwing up tufts of leaves at intervals. The underground stems hold the sand together and more sand accumulates round the tufts of leaves. The sedge thus plays an important part in building up the dunes. Marram grass (*Ammophila arenaria*) first appears on slightly older dunes and its underground stems again hold the sand together. It forms large clumps which trap more sand, and dunes colonized by marram become quite substantial hills. It differs from the other grasses because the leaf margins are inrolled, so that the leaves appear to be circular in cross-section.

Other plants found on young dunes, generally at the drift line, include a number with deep tap roots. The horned poppy (*Glaucium flavum*) is a striking example. It has beautifully silky yellow flowers and, unlike other poppies, it has an enormously elongated seed pod. The curled dock (*Rumex crispus*) is common on the seashore as well

as inland. Here it is shown bearing ripe fruits. A particularly handsome plant is the sea holly (*Eryngium maritimum*). In Elizabethan times its tap root was candied with sugar and orange-flower water to make a sweetmeat. As late as the 1860s this

Shining cranesbill, *Geranium lucidum*

speciality was sold at Colchester, by which time the plant had been almost exterminated along the Essex coast.

Among the creeping plants that are early colonizers of sand dunes are sea bindweed (*Calystegia soldanella*) and prickly saltwort (*Salsola kali*), both of which have rather fleshy leaves. It may be that this fleshiness, as well as being a reserve against dehydration, had something to do with the large amounts of salt in the soil. Saltwort certainly accumulates a considerable amount of sodium carbonate in its leaves; in the Middle Ages it was collected and burnt to provide washing soda.

With time, humus accumulates in the sand and

this can be seen in the contrast between the yellowish colouring of new dunes and the greyish colouring of older dunes. The wider range of plants these older dunes support includes other grass species, such as lyme grass (*Leymus arenarius*) and sand couch (*Elymus farctus*). The latter is a close relative of twitch or couch grass (*Elymus repens*) which can be such a nuisance in cultivated ground because of its creeping underground stems.

The two spurges shown here are also found on older dunes but they have limited distribution. Portland spurge (*Euphorbia portlandica*) grows on the south and west coast of Britain and sea spurge (*Euphorbia paralias*), a little more widespread, is found as well on the east coast as far north as Norfolk.

A number of attractive small flowers grow on older dunes. These include several cranesbills, for instance *Geranium lucidum*, and the storksbill (*Erodium cicutarium*). The wild pansy (*Viola tricolor* subspecies *curtisii*) is rather variable in its flower colour— sometimes yellow as illustrated in the painting and sometimes bluish violet, as illustrated on this page.

Old dunes are quite highly calcareous because of the broken fragments of seashell that are mixed with the sand. This accounts for the presence of such plants as carline thistle (*Carlina vulgaris*). Common centaury (*Centaurium erythraea*) also prefers calcareous habitats and is common throughout the British Isles, especially in the south and near the sea. Another

Wild pansy
Viola tricolor

Dwarf centaury, *Centaurium capitatum*

species, *Centaurium capitatum*, is much more local and only found in England and Wales.

On the oldest sand dunes shrubs follow herbaceous plants as colonizers. A plant of calcareous soils that is often found on relatively stabilized dunes is the burnet rose (*Rosa pimpinellifolia*). It is one of the more easily recognized of our wild rose species, with its small leaflets, spiny stems and very dark hips. Sea buckthorn (*Hippophae rhamnoides*) is a native shrub on sand dunes in the south and west of England from Sussex to Norfolk. However, it is so effective at stabilizing dunes that it has been widely planted elsewhere.

Nowadays as leisure increases and travel becomes easier, more and more people take their holidays at the coast, and so disturbance of natural maritime habitats also increases. Sandy beaches are especially attractive to holidaymakers, and sand dune plant communities suffer correspondingly. Damage to the foredunes can cause them to collapse, and the older dunes then become inundated by the sea. This can be combated by erecting fences at danger points and planting marram grass. It is damage to the older dunes by people approaching the beach and by camping that is probably more serious. The natural history interest is greatest in the old dunes and in the wet places, called dune slacks, that sometimes form amongst them, and these habitats should be protected.

Chalk Downs

1   Upright Brome, *Bromus erectus*
2   Crested Hair-Grass, *Koeleria macrantha*
3   Oxeye Daisy, *Leucanthemum vulgare*
4   Tor-Grass, *Brachypodium pinnatum*
5   Small Scabious, *Scabiosa columbaria*
6   Lesser Burnet-Saxifrage, *Pimpinella saxifraga*
7   Greater Knapweed, *Centaurea scabiosa*
8   Glaucous Sedge, *Carex flacca*
9   Sheep's Fescue, *Festuca ovina*
10 Salad Burnet, *Poterium sanguisorba*
11 Dropwort, *Filipendula vulgaris*
12 Clustered Bellflower, *Campanula glomerata*
13 Kidney-Vetch, *Anthyllis vulneraria*
14 Common Spotted-Orchid, *Dactylorhiza fuchsii*
15 Fragrant Orchid, *Gymnadenia conopsea*
16 Quaking-Grass, *Briza media*
17 Yellow-Wort, *Blackstonia perfoliata*
18 Pyramidal Orchid, *Anacamptis pyramidalis*
19 Cowslip, *Primula veris*
20 Hairy Violet, *Viola hirta*
21 *Homalothecium lutescens*
22 Common Milkwort, *Polygala vulgaris*
23 Horseshoe-Vetch, *Hippocrepis comosa*
24 Birdsfoot-Trefoil, *Lotus corniculatus*
25 Selfheal, *Prunella vulgaris*
26 Bee Orchid, *Ophrys apifera*
27 Dwarf Thistle, *Cirsium acaule*
28 Carline Thistle, *Carlina vulgaris*
29 *Pseudoscleropodium purum*
30 Autumn Gentian, *Gentianella amarella*
31 Wild Strawberry, *Fragaria vesca*
32 Fairy Flax, *Linum catharticum*
33 *Ctenidium molluscum*
34 Common Rock-Rose, *Helianthemum chamaecistus*
35 Wild Thyme, *Thymus drucei*
36 Eyebright, *Euphrasia officinalis*
37 Hoary Plantain, *Plantago media*
38 Squinancy wort, *Asperula cynanchica*
39 *Cladonia furcata*
40 *Cladonia pocillum*

Chalk Downs

People have been living amongst the chalk country of southern England for longer than in any other part of the British Isles. The grasslands of the chalk hills are rich in plant species and attractive flowers are to be seen throughout the year—from the hairy violet (*Viola hirta*) in spring to the autumn gentian (*Gentianella amarella*).

The first human settlers to arrive probably found a continuous tract of woodland, which would have been mostly beech. Beech grows well on chalk; oak, on the other hand, requires deeper soil to reach its full development. The settlers cleared this woodland in order to grow arable crops and especially to provide grazing for cattle and sheep. They built a number of great defensive earthworks such as Maiden Castle in Dorset, which were also used for confining flocks and herds at round-up time. Such an earthwork is shown in the background of the painting. Later, in Norman times, grazing pressure on these grasslands was greatly increased with the introduction of rabbits as a supply of easily caught meat.

The downland turf consists of sheep's fescue grass (*Festuca ovina*), other grasses and many other plants. Some of them, such as the plantain (*Plantago media*) have their leaves in a rosette at ground level. In short turf the leaves lie flat and are not eaten by animals.

An intriguing relationship between animal and plant on chalk grassland occurs in the ways that the flowers of orchids are adapted for pollination by different kinds of insects. The spotted orchid (*Dactylorhiza fuchsii*) is visited by bumble bees and large flies. In the fragrant orchid (*Gymnadenia conopsea*) nectar is contained in a long spur and can only be reached by long-tongued moths that fly at night. The pyramidal orchid (*Anacamptis pyramidalis*) also has a spur, but it is visited by butterflies in the day time, as well as by moths. The lip of the flower of the bee orchid (*Ophrys apifera*) resembles a large bee, and may at one time have been pollinated by an insect that does not live in England, for the flowers pollinate themselves. On the other hand, the fly orchid (*Ophrys insectifera*), in which the lip of the flower resembles a two-winged fly with a bright metallic blue patch of colour on each wing, is visited and pollinated by flies.

Apart from their biological interest, orchids are attractive flowers. But many other chalk downland flowers are equally appealing, for instance cowslip (*Primula veris*) which has to be looked for in the damper places, and clustered bellflower (*Campanula glomerata*) which is locally common on most chalk hills. There is much beauty in the small flowers too. Barbara Nicholson has shown common milkwort (*Polygala vulgaris*) in its three colour forms—blue, white and pink. Bird's foot trefoil (*Lotus corniculatus*) has conspicuous bright yellow flowers which are called 'Lady's Shoes' in some parts of England. The seed pods are black and end in a claw; there are usually four (sometimes five or six) in a head, and hence the name 'bird's foot'. Horseshoe vetch (*Hippocrepis comosa*), with somewhat similar yellow flowers, has pods which break up into three or four (sometimes five or six) horse-shoe shaped segments. In kidney vetch (*Anthyllis vulneraria*) the pods are small and round. The un-

Fly orchid, *Ophrys insectifera*

58

usual flowers, which are often called 'Lady's fingers', are very variable in colour. A crimson flowered form is often found near the sea. The scientific name refers to its former medicinal use as a 'vulnerary', that is, a constituent of an ointment thought to aid the healing of wounds.

Selfheal (*Prunella vulgaris*) is another downland flower that used to be considered a useful herb for treating cuts and grazes. Its curative properties were also thought to extend to the inflammation of the throat called 'quinsy'. As its names show, though, squinancy wort (*Asperula cynanchia*) was the preferred herb for treating these kinds of infection. It is quite a delicate plant, and herbalists may have found it difficult to collect sufficient to prepare an effective medicine! Apparently they were not deterred by such drawbacks, for another delicate plant, fairy flax (*Linum catharticum*) was, according to Thomas Johnson in the second (1633) edition of Gerard's *Herbal*, crushed and warmed with white wine for use as a laxative.

Open patches in the turf provide habitats for lichens and mosses. The subspecies of the lichen *Cladonia furcata* that grows on chalk and the related species *Cladonia pocillum* cannot survive too much competition from other plants. The moss *Ctenidium molluscum* also grows in open turf; its neatly arranged branches make it look like a group of dwarf and flattened Christmas trees. It is interesting that there is a strain of this moss that grows on acid soils.

Wayfaring tree, *Viburnum lantana*

The variety of plant species in chalk grassland depends on the continual cutting back of the more vigorous plants by regular grazing. Now that rabbits are less common than they used to be because of disease, closely grazed chalk grassland is becoming scarce, so some plants become rare, and other changes take place as well. As grazing diminishes, tor grass (*Brachypodium pinnatum*) no longer grows in isolated clumps but spreads to cover large areas, its vigorous growth crowding out other plants. If seedlings of shrubs such as hawthorn (*Crataegus monogyna*) and wayfaring tree (*Viburnun lantana*) are not eaten off when they are very young, their growth into bushes transforms grassland into scrub. Scrubland cannot be converted back into grassland simply by increasing grazing pressure; the scrub must first be cleared away. If, on the other hand, scrubland is left long enough, it will ultimately change to woodland, though it may take upwards of a hundred years for tree saplings to become established.

Changes in farming practice are also posing a serious threat to these plant communities. Even though it is only with the massive and regular application of artificial fertilizers that these thin chalk soils remain productive, more and more grazing land is being ploughed up for arable farming. Exploitation of this kind and neglect of old grazings have already greatly reduced the number of localities where the plants in this painting may be found. Ensuring the survival of the exceptionally beautiful plant communities of chalk grassland faces conservationists with a major management problem.

59

Meadows

1 Sweet Vernal-Grass, *Anthoxanthum odoratum*
2 Marsh Horsetail, *Equisetum palustre*
3 Common Sorrel, *Rumex acetosa*
4 Ragged Robin, *Lychnis flos-cuculi*
5 Floating Sweet-Grass, *Glyceria fluitans*
6 Meadow Buttercup, *Ranunculus acris*
7 Devilsbit Scabious, *Succisa pratensis*
8 Marsh Thistle, *Cirsium palustre*
9 Creeping Bent, *Agrostis stolonifera*
10 Meadowsweet, *Filipendula ulmaria*
11 Sharp-flowered Rush, *Juncus acutiflorus*
12 Soft Rush, *Juncus effusus*
13 Sneezewort, *Achillea ptarmica*
14 Bugle, *Ajuga reptans*
15 Marsh Foxtail, *Alopecurus geniculatus*
16 Yarrow, *Achillea millefolium*
17 Yellow-Rattle, *Rhinanthus minor*
18 Lesser Spearwort, *Ranunculus flammula*
19 Creeping Buttercup, *Ranunculus repens*
20 Meadow Vetchling, *Lathyrus pratensis*
21 Hard Rush, *Juncus inflexus*
22 Marsh Ragwort, *Senecio aquaticus*
23 Common Fleabane, *Pulicaria dysenterica*
24 Marsh-Marigold, *Caltha palustris*
25 Cuckoo-flower, *Cardamine pratensis*
26 *Calliergon cuspidatum*
27 Meadow Saxifrage, *Saxifraga granulata*
28 Bog Stitchwort, *Stellaria alsine*
29 *Rhytidiadelphus squarrosus*
30 Southern Marsh-Orchid, *Dactylorhiza praetermissa*
31 Greater Birdsfoot-Trefoil, *Lotus uliginosus*
32 Tufted Forget-Me-Not, *Myosotis caespitosa*
33 Common Marsh-Bedstraw, *Galium palustre*
34 Skullcap, *Scutellaria galericulata*
35 *Lophocolea bidentata*

Meadows

Barbara Nicholson's painting of a water meadow shows the lushness of the plant community that develops in wet grassy places. During late winter and early spring the meadows are very wet and often flooded; in the summer they are used for grazing cattle. Grazing, or possibly cutting the grass for hay, keeps back the growth of vigorous plants that would otherwise swamp the weaker-growing species. However, the balance is a delicate one; if the grazing were too heavy the less vigorous species would be eliminated.

The background shows a wet hollow a little way into the field, and there is an upward slope so that it is drier towards the back. The plants illustrated require different amounts of moisture to achieve their best growth. Floating sweet grass (*Glyceria fluitans*) usually grows in water and may even be found in slow-flowing shallow streams. Marsh foxtail (*Alopecurus geniculatus*) is often found growing in mud at the side of ponds and thus seems to prefer its roots to be in water. Sweet vernal grass (*Anthoxanthum odorata*), on the other hand, grows on drier ground and is very widespread on both acid and basic soils. It is an early-flowering grass and contains a substance called coumarin that gives new-mown hay its characteristic odour. Creeping bent grass (*Agrostis stolonifera*) is another grass of drier ground. It is found everywhere in every kind of grassland and also in waste places.

A plant of particular interest that is shown here is the yellow rattle (*Rhinanthus minor*). The common name refers to the seed capsules, which can be seen below the flower about one third of the way down the stem. They are round flattened pods which contain a few large seeds. When ripe the seeds become loose in the pods and can be made to rattle about inside by shaking the stems. The plant has no proper root system but attaches itself to the roots of grasses and obtains its supply of water and minerals from them. Yellow rattle, like all the plants in this painting, grows on neutral or base-rich soils. A related species, lousewort (*Pedicularis sylvatica*), also partially parasitic on the roots of grasses, is found on wet acid soils.

Lousewort, *Pedicularis sylvatica*

Buttercups have different moisture requirements. The one that is most tolerant of wet conditions is the creeping buttercup (*Ranunculus repens*). It does well on heavy soils, often forming quite large colonies by means of creeping runners. Meadow buttercup (*Ranunculus acris*) requires less wet soils. It can easily be distinguished from creeping buttercup because it does not produce runners and also because there are vertical grooves in the stem. The bulbous buttercup (*Ranunculus bulbosus*) is not shown because it requires the driest soil of the three. The base of the stem is markedly swollen just below the ground and the sepals of the flowers are folded back. Although requiring differing amounts of moisture, all three of these butter-cups are often found in the same field. Spearwort (*Ranunculus flammula*) and a related species, marsh marigold (*Caltha palustris*) usually grow in very wet places, and are often found in ditches. Marsh ragwort (*Senecio aquaticus*) grows in similar habitats.

Another plant of very wet places is lady's smock or cuckoo flower (*Cardamine pratensis*). Several other spring flowers have names associated with the cuckoo. The scientific name of ragged robin (*Lychnis flos-cuculi*) translates as cuckoo flower. The plant with more English names than any other, 'Lords and Ladies' (*Arum maculatum*; over a hundred English names are given for this species by Cecil Prime in his book *Lords and Ladies*) is often called cuckoo pint. Perhaps these cuckoo names were given because the flowers were thought to herald the return of the cuckoo in spring.

One of the rarest plants to be found in meadows is the snakeshead fritillary (*Fritillaria meleagris*), which is now found only in a few places in the upper Thames valley and in one place in Suffolk. Much commoner is the southern marsh orchid, (*Dactylorhiza praetermissa*). It can usually be distinguished from the spotted orchid (*Dactylorhiza fuchsii*)—shown in the Chalk Downs painting—because the leaves are plain green, but sometimes individuals are found with the leaves marked with ring spots. Wherever the two species grow close to one another abundant hybrids are produced. Often they outnumber the parents and are more robust than either.

Meadows are of very considerable natural history interest because of the great variety of plants that grow in them. This richness of species is well illustrated in Barbara Nicholson's painting, and that is why the picture was chosen for the cover of this book. As well as the widespread common species, rarities are also often found—different ones in different parts of the country. Yet the community depends for its survival upon regular flooding in the early part of the year, and regular grazing at just the right intensity later on. These conditions fitted in with a system of agriculture that grew up slowly over a period of some hun-

Marsh ragwort
Senecio aquaticus

Snakeshead fritillary
Fritillaria meleagris

dreds of years and was still operated very widely until quite recently.

Even though the fertility of meadows is largely self-renewing and they require little management apart from control of grazing, they are no longer economically so profitable. Increasingly, they are being drained and converted to drier pasture, which supports only a few species, or to arable fields, in which few wild plants can maintain a foothold at all.

Hedgerows

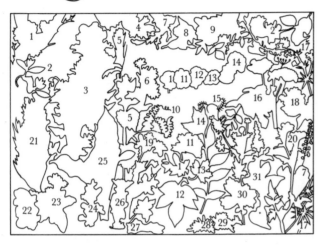

1   Blackthorn, *Prunus spinosa*
2   Honeysuckle, *Lonicera periclymenum*
3   Cow Parsley, *Anthriscus sylvestris*
4   White Bryony, *Bryonia dioica*
5   Garlic Mustard, *Alliaria petiolata*
6   Red Campion, *Silene dioica*
7   Black Bryony, *Tamus communis*
8   Hedge Bindweed, *Calystegia sepium*
9   Dog Rose, *Rosa canina*
10 Bush Vetch, *Vicia sepium*
11 Hawthorn, *Crataegus monogyna*
12 Elder, *Sambucus nigra*
13 Field Maple, *Acer campestre*
14 Ash, *Fraxinus excelsior*
15 Tufted Vetch, *Vicia cracca*
16 Rough Chervil, *Chaerophyllum temulentum*
17 Cleavers, *Galium aparine*
18 Bramble, *Rubus fruticosus*
19 Herb-Robert, *Geranium robertianum*
20 Upright Hedge-Parsley, *Torilis japonica*
21 Greater Stitchwort, *Stellaria holostea*
22 Lesser Celandine, *Ranunculus ficaria*
23 Primrose, *Primula vulgaris*
24 Common Dog-Violet, *Viola riviniana*
25 Lords-and-Ladies, *Arum maculatum*
26 Wild Strawberry, *Fragaria vesca*
27 *Brachythecium rutabulum*
28 *Lophocolea cuspidata*
29 *Plagiothecium denticulatum*
30 Wood Avens, *Geum urbanum*
31 Hedge Woundwort, *Stachys sylvatica*

Hedgerows

Hedges have been part of the British countryside at least since Saxon times, but the extent of hedgerows has varied greatly from one period to another. In medieval times, for instance, farming was largely carried out on an open-field system and hedges were comparatively few. With improvements in farming, particularly in the breeding of animals, land was enclosed in order to control the movement of stock. Hedges reached their maximum extent in the nineteenth century with the culmination of the enclosure movement.

Now that arable farming is increasing at the expense of stock rearing, hedges are disappearing again. Many are being uprooted in order to make large fields and others are being replaced by fencing because of the cost of skilled labour to carry out the essential maintenance that hedges need.

Some of the earliest hedges may have been formed by leaving the margin of a woodland when the rest of it was clear-felled, but most have been planted. The hedging plant that has generally been used is hawthorn (*Crataegus monogyna*). It is a vigorous, light-demanding shrub that readily colonizes open ground and, if left to itself, soon forms thickets. Another species (*Crataegus laevigata*), which looks very similar to ordinary hawthorn, is more shade tolerant and is usually found growing singly in woods. The two species can hybridize and hybrids are sometimes found in hedges.

The traditional way of enclosing a field in which cattle are to graze was to put a rail fence around it and plant the cuttings of hawthorn, called quicks, behind the rails. Nowadays wire, often barbed, is used in place of rails. In the past, if there was to be pasture on both sides which might be used for raising bullocks, two especially stout rail fences were erected with the quicks in between. The result was called a double oxer. The hawthorns were allowed to grow freely and in a few years they would reach a height of about two metres. Such a

Field maple, *Acer campestre*

hedge was called a bullfinch, because many small birds would nest in it. At this stage it would be trimmed to about one metre high and it would also be layered; that is, some long stems were bent over and interwoven with the upright ones to form a more impenetrable barrier.

Sometimes specimen trees were left to grow up freely when the hedge had been cut. They provided shade for cattle as well as useful timber. Most people think they improve the general appearance of the landscape. Field maple (*Acer campestre*) is a tree that was frequently used in this way.

A number of plants, including trees and shrubs, become established in hedges because fruit-eating

White bryony, *Bryonia dioica*

birds perch in them and void seeds in their droppings. Elder (*Sambucus nigra*) is a good example of a shrub spread in this way. A number of other plants that appear early in the life of a hedge and that are bird-sown are scrambling brambles and roses. The small-leaved, pink-flowered *Rubus ulmifolius* is the commonest of the brambles; it is unusual in that it will grow on all types of soil, including chalk and heavy clay. There are many other species of bramble which are difficult to distinguish so they are often included under the group name *Rubus fruticosus*. The plant many people consider the most beautiful of all British flowers, the dog rose (*Rosa canina*), is another scrambler. Its fast-growing stems are rather weak but they retain their position in the hedge because their curved prickles catch on to other plants. Honeysuckle (*Lonicera periclymenum*), on the other hand, climbs by twining its stem round other plants, and white bryony (*Bryonia dioica*) has special sensitive short stems, called tendrils, which attach themselves to suitable supports.

Thus the older a hedge becomes the more species of plants are likely to be found in it. It has been estimated that in a 25-metre stretch of hedge, one new species will appear every hundred years. Therefore, assuming that only one species was planted initially, the hedge in the centre of Barbara Nicholson's painting might be 600 years old.

At the bottom of a hedge the plants in flower vary with the season. Spring flowers such as stitchwort (*Stellaria holostea*) are followed by herb Robert (*Geranium robertianum*) and vetches (*Vicia sepium* and *Vicia cracca*) in the early summer and by plants like hedge woundwort (*Stachys sylvatica*) later in the year. Among the most attractive spring flowers are the primrose (*Primula vulgaris*) and the celandine (*Ranunculus ficaria*). A particularly interesting seasonal succession is provided by members of the carrot family (Apiaceae or Umbelliferae), an easily recognized group, in which the stalks of the flowers all come from a single point like the ribs of an umbrella. The first to flower in southern England is cow parsley (*Anthriscus sylvestris*), but in the north it is sweet Cicely (*Myrrhis odorata*). Later in the season comes chervil (*Chaerophyllum temulentum*) followed by hedge parsley (*Torilis japonica*).

Hedgerows can therefore be seen to have quite a complex structure consisting of several layers. These are the emergent trees, the shrub layer, the scrambling and climbing plants and the herbaceous layer. Hedges are also an important habitat for a great variety of birds and small animals. However, managed hedges require maintenance at least once a year, consisting of trimming and layering, and mowing at the base of the hedge. The skilled tasks involved cannot be carried out entirely satisfactorily using machines, and this makes maintenance expensive for the landowner. Yet it would be a very great pity if this once familiar aspect of southern England, associated particularly perhaps with nineteenth-century Leicestershire, were to disappear completely.

The background of this painting shows rolling countryside with a pattern of not very large fields separated by hedgerows. Not a single arable field is to be seen; the artist must have been thinking of the years between the wars, for putting whole countryside down to grass only occurs in times of agricultural depression.

Waste Ground

1. Common Nettle, *Urtica dioica*
2. Cocksfoot, *Dactylis glomerata*
3. Common Couch, *Elymus repens*
4. Creeping Thistle, *Cirsium arvense*
5. American Willowherb, *Epilobium ciliatum*
6. Common Poppy, *Papaver rhoeas*
7. Smooth Sowthistle, *Sonchus oleraceus*
8. Rosebay Willowherb, *Epilobium angustifolium*
9. Curled Dock, *Rumex crispus*
10. Mugwort, *Artemisia vulgaris*
11. Coltsfoot, *Tussilago farfara*
12. Creeping Buttercup, *Ranunculus repens*
13. Silverweed, *Potentilla anserina*
14. Creeping Bent, *Agrostis stolonifera*
15. Oxford Ragwort, *Senecio squalidus*
16. Sun Spurge, *Euphorbia helioscopia*
17. Field Penny-Cress, *Thlaspi arvense*
18. Hedge Mustard, *Sisymbrium officinale*
19. Pineapple-Weed, *Matricaria matricarioides*
20. Redshanks, *Polygonum persicaria*
21. Field Bindweed, *Convolvulus arvensis*
22. Scarlet Pimpernel, *Anagallis arvensis*
23. Knotgrass, *Polygonum aviculare*
24. Cleavers, *Galium aparine*
25. Scentless Mayweed, *Matricaria maritima*
26. Common Mouse-Ear, *Cerastium holosteoides*
27. Corn Spurrey, *Spergula arvensis*
28. Common Chickweed, *Stellaria media*
29. Petty Spurge, *Euphorbia peplus*
30. Annual Meadow-Grass, *Poa annua*
31. Groundsel, *Senecio vulgaris*
32. Shepherd's Purse, *Capsella bursa-pastoris*
33. Red Dead-Nettle, *Lamium purpureum*
34. Common Field-Speedwell, *Veronica persica*
35. Fat-Hen, *Chenopodium album*
36. Black Bindweed, *Polygonum convolvulus*

Waste Ground

Derelict cultivated land, spoil heaps from mine workings, and sites of demolished buildings provide habitats for the plants shown in this painting.

Such habitats are usually well drained and the soil may be enriched with lime from old mortar and cement or with humus and minerals such as phosphates from decaying organic material. The scarlet pimpernel (*Anagallis arvensis*) is an example of a plant that requires good drainage and prefers a certain amount of lime in the soil. Stinging nettles (*Urtica dioica*) on the other hand only grow where there is plenty of humus and mineral salts. The plants that colonize these habitats have spread from naturally disturbed ground, such as landslides and other sites of natural disaster, where there is little competition from other plants.

Some of them, called ephemerals, develop very rapidly from germination to flowering and seeding so that they may produce several generations in one season. Seven species are illustrated in the foreground of the painting as seedlings and as mature plants. Ephemerals have a remarkable power of adapting their size to habitat conditions. Annual meadow grass (*Poa annua*) is a good example—it may be no more than a small tuft on a gravel path, but in good soil it can form a substantial clump.

Annuals take a year to complete their life cycle. A good example is red deadnettle (*Lamium purpureum*) which is very common throughout the British Isles. A related species, henbit (*Lamium amplexicaule*) is also common in Britain, although it is only found locally in Ireland. Some annuals have been associated with human activities for a very long time; seeds of the poppy (*Papaver rhoeas*) have been found with barley in storage jars in ancient Egyptian tombs. Others are more recent arrivals. Oxford ragwort (*Senecio squalidus*) was grown in the Oxford Botanic Gardens in the eighteenth century and was first noticed growing on walls in various

Henbit deadnettle, *Lamium amplexicaule*

parts of the city in 1794. The coming of the railways encouraged its further spread; the ballast of the permanent way suited it as well as the lava flows of Mount Etna in Italy, which were its original home. Destruction of buildings by bombing in the 1939–45 war provided further habitats and the plant has remained common in cities as well as on railways ever since.

The perennial colonizers of open ground take longer to become established. They produce abundant seed but also large deep taproots or spreading underground stems. If they are cut down, new growth soon comes up again, so that they survive well on disturbed ground.

Some of them, such as the creeping thistle (*Cirsium arvense*) and the curled dock (*Rumex crispus*), are long established weeds of cultivation, but others are more recent arrivals. Rosebay willowherb (*Epilobium angustifolium*) was not especially common a hundred years ago, but since about 1920 it has been spreading quite rapidly and invading waste places in towns. In the early 1940s it became a conspicuous plant on bomb-damaged sites. Another willowherb (*Epilobium ciliatum*), first appeared in England in 1891 and is still gradually spreading.

Many of these plants have been accidentally spread by human activities. Others, such as mugwort (*Artemisia vulgaris*), which were thought to have medical value, were encouraged in the past. Yet others, such as tansy (*Tanacetum vulgare*),

Tansy, *Tanacetum vulgare*

appear sometimes on waste ground, having been discarded from gardens. In well cultivated land, weeds are becoming scarcer because of the use of purer crop seed and selective herbicides. There will always be some disturbed ground however and none of the plants in Barbara Nicholson's painting are likely to become extinct.

Common poppy, *Papaver rhoeas*

Conservation of Plant Habitats

As can be seen from the paintings in this book, there are still many kinds of wild flowers and other plants to be found in Britain. But even those that are plentiful are less so than they were, and others are becoming scarcer. In most cases the reason for this is that their habitats are being altered or destroyed. How this happens is explained in the accounts of particular habitats, especially heaths and wetlands such as fens.

One of the main causes of these changes is increasing population, which results in more land being used for building towns and more land being needed for growing food crops. Another main cause is improving living standards, dependent upon more factories, more ports for oil tankers, freighters and aeroplanes, and more roads built to more elaborate specifications. It also means that more people use their increased leisure time to travel farther afield so that no part of the world is now free from human interference. Human activities inevitably alter habitats. This can be seen for instance in the Cairngorm mountains in Scotland where ski lifts have been built to promote winter sports. Tourists use them in the summer as well, and the tops of the mountains near the upper ski lift stations have been severely eroded so that large areas are now devoid of vegetation.

Farming and industry use up land, and many of the chemicals used cause pollution of the atmosphere and water supplies, producing habitat alterations far from the immediate site of operation. Almost invariably the changes man causes result in a reduced number of species in the ecosystem. This can be seen when previously wild areas are brought into agricultural use by clearing, draining and ploughing the land. The resulting farmland always has fewer species growing on it than the original woodland, heath or fen. And when farmland is taken over for building factories or houses the resulting town has fewer species still. Moreover, these changes are always in the same direction, and cannot readily be changed back again. Villages that were abandoned in the Middle Ages after the Black Death are still easily traced in the crops growing where they stood even though the ground has been ploughed year after year for centuries; the crops do not grow so well where the houses were. Still less is it possible to recreate good farmland out of derelict sites, or where town refuse has been deposited and covered over with soil. Similarly, abandoned farmland does not change back into the type of ecosystem that it was before. Maybe heathland, wetland and woodland species would re-establish themselves eventually, but it would take many hundreds, perhaps thousands, of years.

The impoverishment of regenerated ecosystems can well be seen when virgin forest is clear-felled and then abandoned. The secondary forest that develops has far fewer species in it than the ecosystem that has been destroyed. Of course, this is not really surprising because it took millions of years for the primeval forest to reach the state of balance that it had achieved before human activities destroyed it. Most of the natural forests of Britain and Europe were destroyed long ago, and many of those of North America rather more recently. Now it is the turn of the tropical forests of South America, Africa and South-East Asia. Already half the tropical forests of the world have been destroyed, and further destruction goes on at an alarming rate. It seems likely that by the end of

Wistman's Wood, a relict oak wood on Dartmoor.

the century only remote remnants on steep slopes or terrain daunting even to modern machinery will survive. The forest is disappearing more rapidly than we can study it and learn about it, but we know enough to realize that it is the richest ecosystem that has ever existed on the earth. Thousands of species are being driven to extinction by our relentless and often wasteful exploitation of the world's most valuable resources.

The few bright aspects of this gloomy picture are faint indeed. There is an International Union for the Conservation of Nature and Natural Resources based in Switzerland, which is working to prevent the total destruction of the natural vegetation of the world and the wild animals that depend upon it. The role of an organization such as I.U.C.N. can only be fact-finding and advisory. Positive action based on these facts and advice will have to come from governments. In turn, governments can only be deflected from pursuing short-sighted and short-term ends by informed public opinion. Already some governments are promoting conservation by setting up National Parks in which exploitation and destruction of ecosystems is prohibited by law. Sadly, many of their citizens do not realize the importance of these provisions. Conservation to be effective involves the expenditure of large sums of money, and governments can only divert such sums to causes that have full and convinced public support.

In Britain and other highly-developed countries, habitat destruction is less spectacular and more insidious. We should in future try to ensure that no more farmland is taken for building until the large areas of derelict urban land that now exist especially in our large cities have been fully-redeveloped. Similarly no more wild landscape should be converted to farming until all existing farmland is being properly utilized. This would give us the necessary breathing space, perhaps into the next century, by which time our population might well be stabilized and further inroads on natural habitats would no longer be necessary.

Furthermore, Britain as a world power should make every effort to encourage nature conservation in the rest of the world by providing practical help not only in the form of expertise and advice but also by means of financial grants and loans to poorer countries to enable them to preserve their natural heritage which should in fact be ours as well. The ordinary person can play a part by joining and supporting a local County Nature Conservation Trust and by contributing to the World Wildlife Fund, which gives financial support to many projects initiated by I.U.C.N. Above all, you can help by trying to show others, especially young people, the beauty and interest of natural ecosystems so that they come to realize that it is well worth the effort – the considerable effort – necessary to ensure the survival of at least some of them. This is something that Barbara Nicholson, with some considerable success, devoted the latter part of her life to doing.

Index of Plants Illustrated

Acknowledgements

The publishers are grateful to the British Museum
(Natural History) for providing transparencies of
the original paintings featured in this book

The publishers are grateful to the Oxford
University Press for their kind permission to
reproduce the plant illustrations by Barbara
Nicholson, which appear on the following pages of
this book:

31(L), 34, 39, 42, 43(2), 46(2), 50, 51,
54, 55(2), 58, 62, 63(2), 67, 70, 71(R)
from *The Oxford Book of Wild Flowers*
(OUP, 1960)

19, 22, 23(2), 26, 27, 35(2), 38 from
The Oxford Book of Flowerless Plants
(OUP, 1966)

30, 31(R), 71(L) from *The Oxford Book
of Food Plants* (OUP, 1969)

18, 59, 66, 73 from *The Oxford Book of
Trees* (OUP, 1975)

The publishers are grateful to Jane Coper
for providing the photograph of her mother
on page 10.

Phototypeset in Baskerville by Tradespools Ltd.,
Frome, Somerset
Colour reproduction by J. T. Graphics, London
Key line drawings by Jane Herridge

Designed by Enid Fairhead and Caroline Dewing
Edited by Robin L. K. Wood